Praise for

"*Backbone* is an appropriate title for Sonn... personal, professional, and career reflections. As this fascinating medical memoir so aptly illustrates, having "backbone" implies unwavering personal integrity, strength, and commitment in an individual. The patient management experiences and professional vignettes he offers in Backbone are worthy of thoughtful reflection."
—Mark N. Hadley, MD
    Professor of Neurological Surgery and Director of Residency Training, University of Alabama

"It's rare to read a book about neurosurgery that you can't put down. But this saga of Dr. Sonntag's rise from youthful German immigrant during the Cold War to premier neurosurgeon at Barrow Neurological Institute is spellbinding. From the technical details of surgeries to the political turf wars within the hallowed walls of a world-class hospital, his fascinating tales of what it takes to succeed in the high-stakes world of medicine describe a human journey that remains hidden to most of us. This book, a beautifully written window into the human side of a top neurosurgeon, is a must read, not only for all pre-med and med students but also for anyone interested in or involved with medicine."
—Suzanne Chudnoff
    Financial Systems Analyst

"*Backbone* is a gripping look at the evolution of spinal neurosurgery, but its real impact goes beyond the medical. Historically interesting, inspiring, sometimes humorous, often touching, it is a story that invites readers to examine the ways in which the author's core values of hope, honesty, hard work and humility have shaped their own lives. It is a book, simply put, that instills in readers the desire to become better human beings themselves, no matter what their circumstances.
—Sr. Sara Marie Belisle, OSF
    Franciscan Sisters of Dillingen

"*Backbone* is the story of the determination and integrity of a German immigrant who, despite the stigma of his fatherland's sins, rose to become one of America's top neurosurgeons. It would inspire any young person in this country to follow Dr. Sonntag's example of hard work and resolve to live the American dream."
—Suzanne Baars, MA, LPC, LMFT, LCDC
    In His Image Christian Counseling

Backbone

To write this book I have relied on my memories—and those of close family, friends, and colleagues—to recreate events, locales, and conversations over many years. I have researched facts when I could, consulted with several people who appear in the book, and referred to written documents. To protect their privacy, I have changed the names, and in some cases the identifying details, of many but not all of the individuals described in this book, and I have used composites of some patients in describing surgical cases. I have done my best to ensure that the story I tell here is a truthful one.

Backbone: The Life and Game-Changing Career of a Spinal Neurosurgeon
Copyright © 2016 by Volker K.H. Sonntag, M.D.
Published by Lisa Hagan Books

Design: Smythtype Design

ISBN-13: 9781945962752
Printed in U.S.A

Backbone

The Life and Game-Changing
Career of a Spinal Neurosurgeon

Volker K. H. Sonntag, M.D.
with Jeanne Belisle Lombardo

LISA HAGAN BOOKS

All things excellent are as difficult as they are rare.

—Baruch Spinoza

CONTENTS

CHAPTER 1

THE MIRACLE BOY

You can't suppress creativity, you can't suppress innovation.
JAMES DALY

AT 7:35 ON A DESERT spring morning in 1989, a series of events unfolded that would firmly position me in the emerging field of spinal neurosurgery. It was at that moment that a hard-charging ten-year-old named TJ set off for school on his bike from the modest house his parents rented in the west Phoenix suburb of Glendale, Arizona. In the following moments, he threaded his way down side streets through a neighborhood of low-slung, cement-block houses until he reached a busy four-lane thoroughfare. Already a stream of traffic surged in tight clusters, in both the north- and southbound lanes. TJ braked on the sidewalk midway between the two traffic lights. He was about to ride farther up the road to the traffic light when a window opened up. On a snap decision, he bolted his bike off the sidewalk and onto the road.

ASU professor of journalism and writer, Ed Sylvester, described the accident, how the last thing TJ saw was the red pickup truck bearing down on him at 45 miles per hour; how the impact crumpled the bike into twisted scrap metal and slammed him over the right fender, before hurtling his body fifteen feet into a heap by the curb. Sylvester also described the two sets of finger streaks swiped in the dust of the fender, front to back, sign of the boy's last-ditch attempt to ward off disaster.

The impact should have killed the boy outright. But emergency crews arrived within minutes, and just before eight o'clock, braced with a soft collar and packed into a pressurized safety suit, he was loaded into an evac helicopter and flown sixteen miles away to Saint Joseph's Hospital. There, TJ's destiny would soon dovetail with mine.

The messenger was a junior resident named Brian Fitzpatrick. Lanky and bespectacled, he was notable for the patch of dark hair sprouting from the V-neck of his scrubs. As we strode down the corridors to the ER, Brian filled me in on the details.

"Accident victim," he said, "a kid, hit by a truck on his way to school. Dr. Schiller and the trauma team have done the initial examination. Pretty bad injuries: deep abdominal lacerations and a compound fracture of the right leg."

"Any X-rays?"

"Yes, they're showing some internal bleeding. And it looks like there's a line of separation between the first vertebra and the skull. One of the paramedics who brought him in said the neck felt loose."

"That's not good," I said, swinging through the doors to the emergency department in time to see a trio of nurses rolling the comatose boy out on a gurney. They'd removed the protective suit and stripped TJ down to his underwear. His bruised face bloomed from the supportive neck of a soft collar. Above his closed eyes, his shaggy brown hair was a tangle of dried blood and dirt.

"Give me a minute with him," I said to the nurses.

Bloody and battered, the left side of the boy's body was rigid. Thin tubes sprouted from his bandaged wrists and stubby nose. I leaned over him, dug my fist into the shallow cavity of his chest and rubbed vigorously against his bony sternum. That should have hurt like hell to anyone awake. His face remained passive, but his right leg jerked—definitely a purposeful movement.

I walked over to where trauma surgeon William Schiller and chief resident Tom Graham stood peering at the X-rays on the lighted wall panels.

"So what have we got here?" I said.

"Damage to the internal organs. Internal bleeding. And that dislocation at the occipital-cervical juncture," Schiller said, indicating the apparent separation of the skull from the spine. "There didn't appear to be any neurological damage at first. Both pupils reacted to light, both sides of his body were responsive to pain, he was moaning. The accident report noted some decerebrate posturing at the scene, but that had stopped by the time he got in here. He started lashing out, though, during the examination."

I shook my head. The decerebrate posturing—a rigid extension of the body and limbs, often accompanied by a corkscrewing of the hands and arms—almost certainly indicated injury to the brain, as did the lashing out.

"With that dislocation at C1, movement like that could either kill him or paralyze him," I said. "We've got to keep him quiet."

"Right. To be safe, we immediately sedated him. We're taking him in for exploratory surgery on the internal injuries once the CT scans come back. The nurses have him in the scanner now."

"And the vitals? Any changes in heartbeat or blood pressure?"

"He's stable," Schiller said, "but something strange happened. The boy went into total paralysis. He came out of it almost immediately, but he seems to be wavering between normalcy and partial paralysis on his right side."

That was another ominous clue to what was happening inside TJ's brain. Paralysis also implied damage to the brain or brain stem. And while damage to any part of the brain is bad news, an injury to the brain stem often spells a death sentence.

"OK then," I said. "I'll want to see the CT scans as soon as possible. Has anyone ordered an MRI?" I asked, turning to Graham.

"Not yet," he said.

"Get the order in stat then," I said, thinking what a lucky break it was for the boy that Barrow had only weeks before begun field-testing an experimental MRI scanner. Capable of knocking out digital watches and pacemakers, the magnets on the old one would also have interfered with TJ's respirator. And the respirator was all that was keeping him alive.

The rest of that morning my fellow Stephen Papadopoulos sprinted films over to my office as they became available. When the CT scans came in, they confirmed a definite dislocation between the first vertebra and the skull. They also prompted more questions. Steve and I peered at the images like homicide detectives trying to force a secret out of a dead man.

"See that shadow along the brain stem?" I said.

Steve nodded. "A hemorrhage."

"Most likely. But the question is, is it still bleeding?"

"There's something not right with the brain stem either," Steve said.

"I see that. It's pushed back. There's definitely something else going on there."

I fit in my outpatients and post-op patients between Papadopoulos's updates, feeling like a horse straining at the bit. The new scanner would provide the detail we needed, but it was excruciatingly slow at building the images. Finally, in mid-afternoon it sputtered out a decent set of images of TJ's brain and spinal cord. A quick look was enough for me to send word to Robert Spetzler, director of BNI and arguably the best neurovascular surgeon in the country, to join me. The neurosurgical team was now on alert.

I knew Spetzler had been in surgery all morning performing a hypothermic cardiac arrest (HCA) on an aneurysm patient. The so-called standstill technique drains all the blood out of the body and brings the temperature down to 60 degrees—thus reducing life-threatening bleeding. HCA was then sci-fi-worthy cutting edge and a grueling procedure. Even so, it wasn't uncommon to move immediately from one surgery to another. Neurosurgery is the extreme sport of the world of medicine. At any rate, Spetzler ran a tight ship and had surely already gotten word of the emergency.

Moments later Spetzler, our pediatric neurosurgical specialist Hal Rekate, Fitzpatrick and I huddled around the lighted imaging panels in the patient examining area. For a moment we were all struck mute. None of us had seen anything like it.

I had accepted the likelihood of a hemorrhage but now the MRI revealed a blood clot the size of a large marble pressing against the boy's brain stem above the back of his throat and in front of the spinal cord. We would have to drain the clot as soon as possible to prevent massive damage—if, that is, the damage hadn't already been done. It had been more than seven hours since the accident.

Even more shocking than the blood clot was the extent of the separation between the skull and the first cervical vertebra. The C1 vertebra itself was remarkably intact, but it looked as if someone had taken a knife and hacked it clean from the skull—there was a 2-centimeter gap between the two structures. The tough supportive ligaments had snapped; only a few muscles, skin, blood vessels, and the delicate spinal cord held the skull in place. God knows how the boy had remained alive. As far as we knew, any patient suffering such an injury had died immediately.

The pediatric neurosurgeon, Hal Rekate, shook his head. A big, affable man with a booming laugh, he spoke now with a sober urgency: "Where do we even begin?" he said.

"It's going to be a challenge," I agreed. "We'll have to drain that clot before we can do anything else, and deal with the hemorrhage. But how are we going to get at it?"

Spetzler spoke up in his typically unruffled baritone voice. "It couldn't be at a worse spot. The most direct way would be through the mouth, of course, but cutting through the meninges at the back of the throat will present another challenge. Judging from the hemorrhage, the meninges may have already ruptured, but if they haven't, I don't think we can risk infection from the bacteria. Once that gets into his brain, it'll kill him in days, if the clot doesn't do the deed first."

Spetzler paused a moment, the blue-white glare from the viewing panels throwing the planes of his face into sharp relief. Then he shot me a look. "Even if we get the clot out of there, Volk, how are you going to reattach the skull?"

I knew Spetzler's mind was racing as fast as my own. Like most neurosurgeons, he was fiercely competitive, and since we'd branched off into our respective subspecialties half a decade before—Spetzler advancing surgeries of the brain and vascular system, and me delving into neurosurgical spine cases—we'd felt ourselves neck and neck in a race to advance neurosurgery through our distinct areas of practice. It was a good-natured rivalry but sharp edged nonetheless; it spurred each of us on.

"That's clearly a severe C1 occiput dislocation," I said. "But anatomically, the spinal cord appears to be intact. There could be damage—we know movement doesn't mean the cord's not injured—but it's not severed, and that's a plus right off."

While I was thinking through the dislocation, Spetzler had come up with his own plan for the clot. "OK, an anterior approach is too risky. We've successfully reached aneurysms at the front of the brain stem by coming through the rear. We can take such a posterior approach now and work around to the front."

Even as Spetzler spoke, the second part of the puzzle fit together in my head. "Fusion," I said. "Using the rod and wires. The same procedure I've done on the rheumatoid arthritis cases. I know it's never been used as the sole link before, but if it could strengthen the occipital-cervical connection on RA patients, it might be worth a try."

Spetzler pulled on one end of his brown mustache. "There's nothing to lose," he said. "If we do nothing, he'll surely die."

We scanned each other's faces. A deep crease cut between Hal Rekate's eyes. "Nothing to lose?" Rekate said. " We might save his life, but what if he comes out a vegetable?"

"That's a possibility," I said. "But we know only a few things for certain here. He's got a mother of a blood clot pressing on his brain stem. He's got an occiput/C1 dislocation. The spinal cord appears intact, but it's had a load of stress on it and it could well be injured. The boy's survived two major injuries that by all measures should have killed him. And he's got limited movement. I say we go for it. These injuries are not going to heal themselves."

I knew then that as impossible as it sounded, I was somehow going to reattach that boy's head to his spine.

It was four o'clock in the afternoon before they brought TJ out of the first surgery. Schilling had managed to save half the boy's spleen and repair damage to the kidney and liver. The fractured leg was set in a cast. But the operation had taken over four hours, plenty of time for the hemorrhage in TJ's brain to wreak more havoc. The neurosurgical team wouldn't know what further damage had been done until we got TJ into surgery. Judging from appearances our efforts seemed increasingly futile. Except for the robotic breathing of the ventilator keeping him alive, the boy was deathly still.

Before we started, I joined TJ's mother, Kerry Mathias, in the waiting room. She was sitting where I had found her after examining the boy that morning. A weary, work-roughened woman in jeans and a rumpled print cotton blouse, she jumped up the moment I came through the doors. Her face was haggard and her hands alternated between pushing strands of long dark-blonde hair out of her eyes and covering her mouth.

I knew Hal Rekate had delivered the dire odds to her mid-morning, telling her that the neurosurgical team had one chance in twenty of saving TJ's life. And that didn't mean TJ would regain consciousness, only that he would live. Now, as the time approached for our shot at TJ, I wanted to make sure she understood what I was about to attempt.

"We're ready to begin surgery on the clot and dislocation," I said. "With the clot, it's a matter of getting to it. Once we're there, it should be a relatively quick procedure to drain it. But it's located in front of the spinal cord, and it's going to be difficult to reach."

I paused to let her take it in. "With the separation at C1—the separation of the skull from the first vertebra in the neck—I'm going to try something new. It's a long shot, but I'll reconstruct the connection with bone from TJ's hip. Then I'm going to put a rod in to support the spine while the bone fuses and strengthens. The fusion should facilitate strong new bone growth over time."

Kerry's face pinched up in doubt.

"Think of it this way," I said. "It's like rebar in cement. The steel rod I'll put in will support his cervical spine until new bone grows."

She nodded. "OK Doctor, do what you can."

Then I added one last, difficult reminder. "You know there's a chance he won't wake up. And there's no guarantee that he'll have movement in his arms and legs if he does wake up. But we have to take this chance. Otherwise, well, it's not a good outlook."

She took a deep breath and exhaled slowly. "I guess we'll take it one day at a time," she said.

I shook her hand and watched as she fumbled in the scuffed bag slung over her shoulder, pulled out a pack of cigarettes, and walked distractedly through the street doors. I hoped I would have better news to give her once it was all over.

Our first task was to remove the soft collar and backboard to which TJ had been strapped since early morning. Enough to stabilize him during surgery on his abdomen, it wouldn't provide enough support to hold his neck stable now.

"We need a halo in here," I called out, referring to the brace used to immobilize the cervical spine.

But then, another frustrating delay. The hospital had no brace that fit a child. Another hour passed as the orthopedics department cut one down to size. The anesthesiologist set the drip of intravenous painkillers going and fit the inhalation mask over TJ's nose and mouth for the anesthesia. The EEG technician pasted half a dozen flat metal electrodes to his scalp. And when the halo finally arrived, Robert and I fitted the lambswool jacket around TJ's narrow chest, aligned the four black steel rods with the halo encircling the top of the boy's head, and bored the screws into his skull that secured the metal crown. Satisfied that his head, neck and chest were utterly rigid, we rolled him over onto his stomach.

It was 6:00 p.m. Robert stepped up and I moved into place across from him to assist. Using a Bovie cauterizing knife, Robert slit a dissect down the midline to the bone, opening the back of the neck from just above the base of the skull to the fourth vertebra. I pinned back the muscles and skin with a self-retaining retractor and started suctioning. As the blood drew away into the suction tube, I saw the opaque bluish gleam of the dura mater covering the spinal cord where it showed through the gap between the skull and the first vertebra. The clot lay below but out of reach, wedged under the spinal cord, itself overlaid by the carapace of that first vertebra.

Robert took a breath. "I think we'll have to get that lamina out of there," he said, referring to the arched plate of vertebral bone at C1. "I don't see how I can get around to the front otherwise."

"Right," I agreed, suctioning the thin stream of blood seeping into the wound.

Robert deftly performed the laminectomy, carving out a neat window of bone and exposing the spinal cord. He gently opened up the dura with an 11-blade knife and together we peeled it back and sutured it to the muscles along the side. Now, our view opened up, the microscope came in.

There was the clot, wedged in under the spinal cord, like a small damn ready to burst. Dark, viscous blood oozed from it in small clumps. But there was no upwelling of bright, new blood, no new hemorrhage that had been held back by the pressure of the clot and set gushing by our very efforts to save the boy.

Robert painstakingly pried his way through the thicket of tissue and nerves, around to the front of the spinal cord. Moments later he reached the jellied, black knot of blood. "Sucker," he said. The surgical nurse slapped a small canister sprouting a narrow tube into his hand. He pierced the clot and quickly drained it. A sizzle with a Malis bipolar coagulator—long, tweezer-like electrosurgical forceps—on the tattered blood vessels that had created the clot and Robert was done.

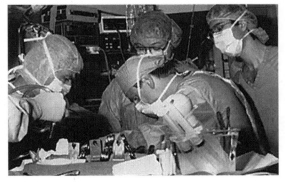

With Hal Rekate and Robert Spetzler during the first phase of TJ's surgery, as resident looks on.
Photograph used with the permission of Barrow Neurological Institute.

I was up. Chief resident Tom Graham scrubbed and moved in to assist me. I'd done the procedure I'd decided on three—perhaps four—times but never on a patient as fragile as TJ. Had I had a night to sleep on it, anxiety would have gnawed at me till dawn, but now a bomb could've gone off and I wouldn't have flinched.

I glanced at the surgical tray where my tools lay: scalpels; a blunt-edged bone chisel; a delicate spoon-like curette for scraping; forceps and scissors; dissector and retractor. To this array I had added several 6-inch-long threaded steel rods and a selection of bendable aluminum skewer-like pins of equal length. Off to one side, a common bench vise from Home Depot sat within easy reach, and next to it a pair of toolbox pliers.

The idea for the method I was about to employ had come to me a few years earlier while working with rheumatoid arthritis patients. RA patients suffer not only from a wrenching distortion of the joints and bones but also a ravaging loss of bone density. Severe immune suppression further complicates treatment. Those RA patients that suffered from degeneration of the cervical vertebra, in particular C1, had begun showing up at my door in the early 80s when my name was getting around as a surgeon who could handle complex cervical spine cases. As with many spine cases at the time, the standard treatment was to place such patients into a halo brace and then lay bone down at the occiput/cervical juncture and hope it would fuse. Unfortunately, such porous bone is hard to fuse, that is, to heal and grow together with the farmed bone to produce new nutrient-rich bone. It's like dry pumice compared to the dense grain of healthy bone.

Having partnered with the orthopedic surgeons, I had become familiar with the threaded rod called a Steinman pin, which they used to reduce femur fractures in traction. The idea came to me: Why not bend the rod and use it for internal stabilization on those RA patients while the bone was healing? And it did help augment the fusion in those patients. I later used the device once or twice on other patients who had a tumor of the cranial cervical junction (which is usually taken out through the mouth) and needed to be stabilized from the back.

Not that everyone would sing praises about yet another use of my innovation. By the late 80s the orthopedic surgeons were getting their noses bent out of shape over my doing any kind of instrumented surgeries. That I was training other neurosurgeons to do it got their hackles up even more. In their minds, instrumentation was their purview. But I'd been taking care of cervical spine cases at Barrow since I'd arrived hungry and obscure a decade earlier. And instrumentation was the future of spine surgery. It was also quite possibly the key to TJ's future as well.

The incision gaped like an open mouth, the raw, meaty sides pinned back by the metal teeth of the retractor. TJ's cervical spine was fully exposed, the 2-centimeter gap at C1 as vacant as the missing rung at the top of a ladder. I picked up a *stilette*, one of the thin, soft aluminum pins, and starting at C4, aligned it with TJ's spine on one side. At the bulge of the occiput, I bent it into a "U" shape with the pliers to conform to the width between the spinal processes at the top of the spine. I then aligned the remaining length of the pin along the other side of the cervical spine and bent the top rounded section to conform to the angle of curvature at the juncture between the base of the skull and the neck. Later a journalist likened the resulting "U" shape to a spoiler on a motorcycle.

I paused to view my handiwork. "That looks pretty good," I said. "This is the angle we need. See that?"

That smooth, flexible *stilette* was just the model, however. I selected one of the threaded steel rods from the surgical tray, fingering the pinched end where I'd cut it to size with a rod cutter. Stepping quickly to the side of the room where the vise sat ready, I grabbed one end of the rigid rod with a towel and secured the other end in the vise. Then I bent the new rod with the pliers to conform to the aluminum model. Having achieved the proper curvature, I put the "U" end in the vice and bent again to get the right incline. A rod that had cost $7.50 was now a sophisticated piece of surgical instrumentation.

While I was engaged with the rod, it was time to farm the bone. Graham had already exposed the curved crest of hip. He set to work with the chisel, tapping and prying a long marrow-rich strip out. Moments later he dangled it at me.

"You think this'll do it, Chief?" he said.

"Get me a little more," I said. We'll need some cancelous bone to squeeze into the facet joints and more to lay over that space at the laminectomy."

I paused. I figured a three-point fixation along the base of the skull would provide the most stability. I took a burr hole drill and quickly bored three holes, the two outside holes conforming to the width of the

"U" bend in the rod. I placed the rod down and positioned it for the next step. For this I needed seven wires, each about twice the length of a twisty used on bread wrappers. These were cut from the same stiff 18-gauge wire found in hardware stores, crudely thick compared to the fine, flimsy wire we use today and dangerously sharp when passing by the spinal cord. But it's all we had. I took one of the wires, looped the end of it into one of the burr holes and threaded it under the bottom of the skull and up the other side, making sure the rod lay between the two ends. When the two loose ends were even, I put a scissor-like snap on the wire to hold it in place, and repeated the process on the second and third holes.

It was time for the sides. The vertebra at C1 where we'd removed the lamina was mostly gone, with only remnants on each side. The nerve roots were too close to those remnants of bone for wiring. We had to anchor the rod to the normal bone below. I nodded to Graham. "OK Tom, take the two on your side over there, and I'll handle the ones on my side. Let's get these other wires in." We threaded the four remaining wires under the sloping edge of the lamina and up the other side at C2 and C3, fixing each one in place with a snap. The loose ends of all seven wires poked out like spaghetti. I adjusted the rod, making sure it was positioned properly between the two loose ends of each wire. "OK, let's snug these wires down," I said. Working rapidly, I began with the skull, removing the snap and twisting the wire tightly around the rod before threading it up again into the burr hole and back down for a final twist round the rod. I repeated the process with the second and third holes, until I had firmly secured the U-curve to the back of the skull. Tom Graham started in on the sides, ensuring at each point that the wire was wound tight around the lamina and deeply buried in the threads of the rod. The rod would never slide out of position now; there would be no vertical migration once we closed it all up.

By now the lights overhead felt like a broiler. Underneath my gown, the heavy protective lead apron that shielded my body from the fluoroscopy rays pressed on me like a weighted heating pad. Rivulets of sweat ran down my neck, chest, and back. I paused a moment as the nurse wiped my brow.

Steady bleeps punctuated the silence. On the monitor, rows of lines crested long and shallow, reflecting normal brain activity. TJ's mother had said her boy was tough; now his body marshaled all its energy into keeping him alive.

TJ's cervical spine was stabilized for the short term, but the bone fusion would determine his long-term prospects. The bone Tom Graham had chiseled out waited in a shallow pan on the surgical tray, seeping raspberry wisps into the saline solution. The nurse fished out a piece and handed it to me. I cut a short length off with scissors, dug at the spongy, cancelous center with the shallow spoon end of the curette, and squeezed the marrow paste into the facet joints between C2 and C3, and between C3 and C4. Then I snipped the rest of the hard cortical bone into small strips, scraping each piece with an awl-like tool called a periosteum until the nutrient-rich marrow bled through. I did the same to the occiput and the lamina and lateral mass at C3 and C4, until I was satisfied that both the receiving bone and donor bone were vitally bloodied. I laid those strips along the lamina and lateral mass on both sides. As a final touch, I snipped a piece of bone the size of a postage stamp and sutured the little flap to the rod on either side of the hole where the lamina had been at C1. Already I imagined the growth of new tissue, fusing the skull back to the spine, reversing the terrible assault of the morning now more than twelve hours in the past.

I stood up from the hunch I'd been frozen into for hours. It was up to TJ now, to his individual and mysterious powers of regeneration, to his being and will, to God or luck. I turned to Graham. "Close him up," I said. "We're done here."

It was a relief to strip off the surgical mask and bonnet, to shrug out of the gown and lug off the heavy apron in the cool and quiet of the scrub room. It was late, near 10:00 p.m. The stillness that pervades a hospital at night resonated with the utter exhaustion that washed over me. I washed my hands, thought about changing into dry scrubs but decided against it. One last task awaited me.

I walked through to the waiting room. Kerry Mathias was holding her lonely vigil where I had left her six hours earlier. She jolted upright when she saw me.

"How is he?" she said, gripping my outstretched hand. Did he do okay?"

"Yes," I said, "the operation went fine. There were no big complications, as far as bleeding or misplacement of bone. His skull has been successfully reattached to his vertebra, so he's no longer in danger of severing his spinal cord by coughing or sudden movement."

She gave a small cry. "How much blood did he lose? Where is he?"

"There was usual blood loss, nothing out of the ordinary. He's going to be in the ICU shortly. He's in the halo brace and he's going to be intubated for some time, but you'll be able to see him soon."

"Do you know how long he'll be in ICU?"

"It's difficult to determine, but at least a week. The operation itself went fine but we don't know what it will bring. Neurologically it will be difficult to examine him any time soon because of the other multiple injuries, and he's going to be comatose for some time. We just have to wait and see."

She nodded. "I guess we'll take it day by day. We'll take whatever happens."

I felt sorry for her, but I was propping my own body up by will alone. I shook her hand in parting, then added: "First thing in the morning I'll come back and see him."

I filled my lungs with the cool night air. The lights of the main building ebbed away behind me. As I walked to my car, I felt flat: not hopeful, not worried. I drove out of the parking garage and up a narrow street before turning onto a main artery. Indian School Road had emptied of traffic. I headed east past the shuttered shops, the muted late-night clubs and brightly lit convenience stores, toward the dark rise of Camelback Mountain. I switched on a soft rock station, changed it to the talking heads on KTAR news. My scrubs were cold and clammy against my skin; I longed to shower and change.

The house was quiet when I pulled into the garage. My wife Lynne was probably asleep by now, as were my young daughter Alissa and her little brother Chris. That was as it should be. I never discussed the day's cases with Lynne. The demands of my career had built a rigid separation of duties for both of us, leaving Lynne to go it alone on the domestic front most of the time.

I padded through from the garage into the house. I knew a part of my mind would be on alert through the night: going over the procedure; waiting for the phone to ring; wondering if I had done all I could. But I was drained. Perhaps sleep would come before I started it all over again at seven the next morning.

CHAPTER 2

SHOWDOWN IN THE OR

Without change there is no innovation, creativity, or incentive for improvement. Those who initiate change will have a better opportunity to manage the change that is inevitable.
WILLIAM POLLARD

IT WAS THE THIN, glassy whine of the drill that got the orthopedist's attention, alerting him to my departure from the usual script. A collegial but curt man a head shorter than I, Dr. S. was partnering with me on a cervical spine surgery that day. Hunched down at the patient's hip with one of my two neurosurgical fellows, Ian Kalfas, he was chiseling out a shard of bone from the back of the ilium, which would be used to fuse the spine later in the surgery. I worked up at the neck where the dislocation was, with my other neurosurgical fellow, Stephen Papadopoulos, across from me. Except for the two exposed areas of his body where we probed and drilled, the patient lay submerged face down in a sea of sterile blue cloth. His head clamped between Mayfield tongs and his eyes taped shut, he was long oblivious to the network of tubes and machines that sighed and sucked and beeped around him.

Steve Papadopoulos had completed the prep on the skewed spine, slicing a 3-inch incision at the back of the neck with a # 15 scalpel, then separating the meaty muscle and tissue from the bone with the hot blade of an electric Bovie knife. He worked down until he reached the thick periosteum membrane covering the bone. A few more minutes of delicate chiseling, and the cervical vertebrae gleamed a pearly white through the bloody mass. Using a scissor-handled, tong-like retractor, Steve dug the tines in, spread the sides of the wound apart, and locked the device in place.

The Bovie had cauterized the raw muscle, searing it like meat in a pan and slowing the flow of blood into the wound. Now Steve took over the suction hose, pecking at the seepage like a scavenging bird while I leaned in. There was the break in the normally interlocking column, the errant facet joint of the seventh vertebrae—C7—riding up on its neighbor C6 on one end, and biting into the spinal cord like a sharp incisor on the other. I put a clamp on the spiny processes of C6 and C7—the small bony knobs that rise up in the middle of the spine—and gently pulled the dislocated facet joint back in line with its neighbors. That did it. C7 slid back under C6, relieving the pinch on the spinal cord. The neurosurgical part of the surgery should have been all but finished.

Normally, having aligned the vertebrae, we would have wrapped some 18-gauge wire around the spinal processes and then turned it over to the orthopedist. We may not have even stayed while the orthopedist completed the fusion, laying the shards of hipbone over the lamina. But today was not a normal day. Today we weren't following the conventional route.

Instead of the wires, Steve and I carefully positioned new instrumentation—two small, steel plates resembling part of a door-hinge package—over the re-aligned section of spine. Dr. S. was still gathering bone and could not see what we were doing amid the press of blue-sheathed bodies, the surgical tray and sterile draping. Once the plates were positioned, Steve held them fixed while, as if penciling in where a nail might go on a wall, I marked with a purple pen where the entry holes for four screws would be drilled.

I paused for a couple of seconds, shot a look at Steve, and took the drill; it was a tool that up to now, in a spine case, had been used exclusively by orthopedists. I quickly roughened up the lamina between the spiny processes—to get it to bleed so it would fuse with the farmed iliac bone. Ordinarily the orthopedist would have handled that step too, not the neurosurgeons, but what Steve and I were about to do made it necessary. Then I quickly drilled two shallow pilot holes on Steve's side, where the lamina flattens out into the thick, horizontal lateral mass. That's as far as I got before the lid flew off the cooker.

At the sound of the drill, Dr. S. jerked his head up, as if he'd been shot.

"What are you doing?" he said.

"We are going to put lateral mass plates in," I responded, looking up.

The strip of his face visible above the mask flushed a bronzy red. He stepped over, shouldering Steve to the side, and peered into the cavity where the spine rose from the tissue like a half-excavated fossil. His forehead was barely half a foot from my nose. I could practically feel the heat radiating off him.

"Why are you doing that?" he said. "Wires work perfect in this kind of case."

"The plates will work better for this patient."

He pivoted away from the table and ripped off his blood-splotched gloves.

"Take me off the record," he said, striding towards the door. "And take me off the op note. I don't want to have anything to do with this case."

The doors whooshed back in, then out, then in, before coming to rest with a little thwump. I looked at Steve. His eyebrows had shot up around his hairline, but he had resumed his place across from me.

"Chiefy," I said, "drill the next two holes. We've got something to accomplish here."

I had arrived at Barrow Neurological Institute at seven o'clock that fall morning. Dodging gurneys and breakfast carts, wheel chairs and I.V. stands, I headed to my office. The muted bouquet of soiled bandages and chlorine wafted above the buffed floors, mixing with the din of voices, ringing phones, elevator doors, footsteps and squeaky wheels.

My day would start as usual: the morning rounds with the typical entourage of med students, residents, and visiting doctors. First, though, I would get a quick report from the two new fellows, Steve Papadopoulos and Ian Kalfas. They would have already picked up the charts, gotten reports from the residents on duty overnight, and prepared for the rounds.

Papadopoulos was long and lean with a tidy cap of dark brown hair. Having taken leave from a faculty position at the University of Michigan only four months earlier, he could come off as rather patrician in his manner, but I had liked him immediately. Kalfas, stocky, muscular, and athletic, had come out to Phoenix just before him. Both young doctors had excellent hands. Both possessed superior judgment in the OR. And both had been enticed west by the brand new fellowship I was offering in spinal neurosurgery.

It was 1988. Until then, there had been no such thing as a spinal surgery specialist. There were the orthopedists, the bone doctors, and there were the neurosurgeons, who handled the brain, spinal cord and nerves. But in the realm of spine surgery, such a distinction was becoming blurred. The most difficult cases involved both bones and nerves.

On that day I sensed something was up when I found the two fellows pacing around outside my office, Steve with X-rays in hand.

"You've got a new case in," he said.

"OK," I said, "let's have it."

"Twenty-five-year-old male. Came in overnight. Involved in a motor vehicle accident. He has a C6-7 facet dislocation. And . . . he's quadriplegic."

We pulled up the X-ray on the view box behind my desk. There, the ghostly ladder of spine and the derailed facet joint. Even as I peered at the screen, I felt my resolve ratchet into place.

It was exactly the kind of spinal cord injury I had been waiting for: a facet dislocation between the sixth and seventh vertebrae, at the base of the neck where the cervical spine meets the thoracic. It was an injury that would lend itself to a new instrumentation technique that had been performed successfully on the spine in Europe but was not yet approved by the FDA for use in the United States. I was ready to try the instrumentation on an "off-label" basis, following the lead of a New York neurosurgeon, Paul Cooper. Cooper had traveled to Europe, seen the procedure performed successfully, and then done his own off-label case. It was his article in the journal Neurosurgery that had piqued my interest in the procedure. Now here was a case practically designed for it.

I looked from Steve to Ian and back again.

"Okay," I said. "What's your plan?"

They knew and I knew what we were going to do—after all, we had been following the development of the procedure for months. But this was part of the training. So I did what I always did with those I mentored—I let them lead.

"We'll need to get the facets relocated," Ian said, "and then do a posterior fusion."

"Right," I said, turning to Steve. "What do you think, Chiefy?"

Chiefy" was the title I used with all the residents and fellows; it removed rank from the equation and relayed the message that, for me, our work was always fundamentally a team endeavor. Now Steve drew himself up and went for it.

"This might be the one for the lateral mass plates. We've studied the procedure for months, discussed the possible problems. We've used models and practiced on cadavers. With the injuries this patient has, it'll reduce the complications post-op. I'm satisfied we can do it safely and properly."

Steve had taken the words out of my mouth.

"Right," I said. "This case is perfect. The dislocation is right at 6-7. We don't have to worry about the vertebral artery. We're halfway home."

Nods all around.

"OK, then." I said, "The cat's out of the barn. See if there's a time open in the OR later on today."

Steve and Ian smiled at my typical fracturing of the English language—a legacy of my German roots. Then we all sucked in a big communal breath.

Nothing was more crucial than finding the right spinal cord injury for our first case using lateral mass plates. If the outcome was not successful, it would put us back months, and worse, provide ammunition for those who opposed the technique. Among those opponents was the entire cohort of orthopedic surgeons at Saint Joseph's Hospital, where Barrow is located. This particular patient was ideal because he was—in the parlance—a "complete/complete." He had neither movement nor sensation below the level of injury. More critically, the wrench—the place where the small, spiny shoehorn of bone called the facet had been yanked out of line with its neighbors—had come between the sixth and seventh vertebrae.

And that was the clincher. Steve and Ian knew it the moment they saw the X-ray. Because the unique thing about the seventh vertebrae is that there is no vertebral artery at 7. The vertebral artery enters higher up, at 6. That was critical because the vertebral artery is the one that courses to the brain. Inserting screws in the spine too close to this critical vessel would be risky. Coming in at C7, we had room to maneuver.

Despite my excitement, my overriding consideration was not my desire—however strong—to try a cutting edge technique on an actual patient. I was convinced that using the mass lateral plates was in this patient's best interest, despite the scant hope of reversing his paralysis. I doubted he could beat the odds that experience told me were working against him, but I fervently hoped that we could salvage the minimal movement remaining in his upper arms, and ease his recovery by not having to put him in a halo brace.

He was a handsome young guy with black hair, his body a heavy, motionless weight under the sheet. When I stopped in on the rounds, a pretty, young woman stood by his side, her knuckles white from gripping the bed rails. I bent down to address him.

"From the X-rays, it looks like your injury is pretty severe," I began. "Your spinal cord is damaged. That's why you can't move your arms or legs."

I'd seen the look in his eyes many times: fear, uncertainty . . . complete vulnerability.

"Hopefully we can help you. We'll need to operate to realign your spine and decompress the spinal cord, give it a chance to recover."

"Is there a chance it will recover?"

"It doesn't look good because, as you know, you have no movement or sensation below the level of your injury. But, I have certainly had patients who, after we realigned the spine, did achieve some recovery. Hopefully you'll be in that category."

He nodded and swallowed, his face chalky in the morning light.

"So," I continued, "we want to use something new on you. We want to try plates rather than wires to stabilize your spine. The device is not yet FDA-approved for this particular surgery—for use in the spine—but in your case I would like to try it. It's called off-label use, and we'll need your consent. All other things aside, if the operation goes as I think it will, we should at least be able to do away with the halo."

I could see that, dire as the patient's situation was, he was relieved he might be spared the immobilization device called a halo brace. And for good reason. It's an unforgivingly rigid vest that straps vise-like around the chest and from which protrude, vertically, four steel rods. The ends of the rods support a circular crown—the halo—of stainless steel (now titanium), which is bolted into the patient's skull with four to six screws. The halo stabilizes the spine by limiting mobility but prolongs the ordeal; it must be endured, non-stop, for eight to twelve weeks while the bone heals.

For an acute quadriplegic, wearing a halo is not only torturous but also dangerous. With none of the muscles in the rib cage—the intercostals—functioning, the only thing driving breathing is the diaphragm, which is controlled from nerves higher up the cervical spine. (As the old saying goes, referring to the mid-level vertebrae in the neck, "3, 4, 5 keeps you alive.") Breathing is labored, and labored breathing often leads to pneumonia. And that's not all. The prolonged bed rest shoots up the morbidity and mortality rate from other complications such as infections, blood clots, and pulmonary embolisms. In-hospital-acquired maladies hover like a specter over many patients, especially those with this kind of spine injury. So if I could at least remove one threat to his recovery—the halo brace—I was determined to do so.

I knew before backing through those OR doors that I was pushing the envelope. I knew the orthopedic community was edgy at the changes afoot. Traditionally orthopedists and neurosurgeons worked together on spine cases like this, but as the neurosurgical community forged inroads into what the orthopedists saw as their territory, resistance mounted. The orthopedists might allow us standard procedures on the cervical spine—the neck—but instrumented surgery, especially on anything further down the spine, was stepping over the line.

I found the growing resistance both disappointing and perplexing. I had been taking all spine trauma cases for some time for the simple reason that only neurosurgeons, not orthopedists, took trauma calls. I had already knocked a few noses out of joint by doing a handful of instrumented anterior fusion cases—going in through the front of the neck to fuse bone to the injured vertebrae. It had been a gradual development, but now so many spine cases were coming my way that my interest in the latest procedures—which meant instrumentation—was skyrocketing.

The truth is—as Steve later put it—I was the tip of a spear already in flight, and I wasn't about to retreat. If my colleagues knew anything about me, it was that I had a strong will to change and improve the field, that I was never happy to accept the status quo. The argument "This is the way it's always been done" held no sway with me.

Still, on that day I wasn't particularly concerned. Once the surgery had been scheduled on our end, the secretary had called the orthopedic desk to coordinate it. The orthopedist who picked up the case, Dr. S., had worked with me many times before. We'd even shot baskets together. My mind was not on the politics brewing.

Knowing the procedure was new, I did what I always did: I went over every detail with my fellows, and the scrub nurse, ahead of time—from the size of the screws and plates to the positioning to who would hold the plates in place versus who would hold the screwdriver, (incidentally, at that time, an ordinary Phillips screwdriver from Sears). The prep included a full alternative plan in the event that the plates didn't fit and we had to fall back on the wires. Even as we backed from the scrub sink into the operating room with our dripping hands in the air, we were still going over the steps:

"Chiefies, you ready to rock n roll?"

They nodded.

"OK, the screws need to go in 30 degrees superior and 30 lateral. No vertebral artery at the 7th. We don't have to worry about that one but we still need to be careful at C6."

Then the scrub nurse was holding open our gowns and gloves and rolling the tray of instruments into place. We closed in round the table with the rest of the crowd: Dr. S., already in place; the anesthesiologist squeezed in near the bank of blinking monitors; a couple of visiting surgeons hovering near the head and craning their necks for a good view; the rotating nurse ready at the perimeter; and a resident or two fitting in where they could. Cords slithered across the floor and hung vine-like from stands. The chill, right-angled space seemed to shrink down to the proportions of a space capsule. We were on.

In the moments following Dr. S's abrupt departure, Steve drilled the two shallow pilot holes remaining on my side. Then, instead of risking it with the electric tool, we bored all four holes down to a depth of 12 centimeters with a hand drill. We positioned the new plates, and took our turns with the screwdriver. The screws found their purchase and spiraled in to a snug fit.

Ian Kalfas was still at the hip, finishing what the orthopedist had left undone. He had already given the sawed-off strip of iliac bone to the scrub nurse, who had put it in saline to preserve the moisture. She now handed it to me. I roughened it up as I had done with the lamina, split it in two, trimmed it down to size, and fitted it over the lamina on either side of the spiny processes. A few minutes later, after the visiting surgeons had jostled forward to eyeball the fused spine, my job was done and the team closed the patient up. It had gone without a hitch.

Later, when the patient was groggy but responsive in recovery, I checked in on him. I did not expect a miracle, and indeed the young man still had no movement or sensation from mid-chest down. I was relieved that he had retained the partial movement in the upper arm— the deltoids and biceps—that he had come in with. It may seem a low aspiration, but while all spinal surgeons hope for improved mobility, we wait with baited breath to see that the surgery has not made the condition worse. That small retention meant he might be able to feed himself eventually, with a fork or spoon taped to his hand.

Steve, Ian, and I were even more relieved the next morning. The post-op X-rays, taken with the patient in both a prone and a sitting position, confirmed that the plates were holding perfectly. There was no instability, no screw pullout or loosening. The new procedure had survived its first challenge; successful fusion was almost guaranteed. Our decision to use the new instrumentation had been vindicated.

Only afterwards, when the word spread about the altercation, did I realize what a watershed that operation was. It wasn't the "shot heard round the world" of spinal surgery that the TJ case would be, just months away, but it did further pit the two camps—the orthopedists and the neurosurgeons—against each other. It was another skirmish in what would soon turn into a pitched battle for control of the spine.

In the larger, purer sphere of advances in medicine, the case also marked a kind of "punctuated equilibrium"—a sudden jump up—in the evolution of neurosurgery. Later, that is what struck me most acutely, the progress—and the attendant hope—that this further development would bring to the individual patients who suffered from some of the most devastating ailments and injuries known to humankind, assaults on the very control centers of their being.

I was intimately acquainted with such maladies and with the limitations of earlier practices in neurosurgery. Some three and a half decades earlier, as a boy of eight in Cold War Germany, I had stood by helplessly as the field I later entered fell short of delivering my father, whole again, back to his family.

CHAPTER 3

THE BRAIN ABSCESS

We look before and after, And pine for what is not;
Our sincerest laughter With some pain is fraught;
Our sweetest songs are those that tell of saddest thought.
PERCY BYSSHE SHELLEY

I DON'T REMEMBER the exact moment that I knew my lifelong mission would be to understand and repair the neural pathways of the human brain and spinal cord, but for as long as I can remember, the image of the scar on my father's right parietal-temporal lobe has haunted me.

The scar is absent in a studio photograph of him that has resided on a shelf in my office for thirty years. That photograph, taken in the 1930s, shows a robust, clean-shaven man smiling under a slim-brimmed, high-crowned fedora hat. He wears a suit and a fat, broadly striped tie. He's a man in charge of his life with a future to look forward to. Marriage. Family. A profession as a dentist, of which he could be proud.

World War II derailed that future for nearly a decade. First came compulsory service with Hitler's army in the East. Then separation from my mother in the chaos of Nazi Germany's collapse. Four grim years in an allied refugee camp followed.

But my father regained his footing in Germany's economic recovery of the 1950s, the *Wirtschaft Wunder*. He reestablished a thriving dental practice in the pretty old spa-town of Bad Hersfeld in central Germany, leaving the upheaval of the war and Germany's defeat behind him. Memories of his mother's fulfillment of a patriotic suicide pact at the end of the war could be repressed under new hopes for the future. He, my mother, brothers and I had survived the war and its aftermath.

He hadn't counted on further disaster, though. Not after all he had experienced. And not just when everything he'd wanted had become reality at last.

The brain abscess announced itself one evening in October of 1952. My parents, two brothers and I sat squeezed around the table in our small kitchen, cutting our Abendbrot (evening meal) on the wooden boards that served as plates. My brothers and I had made it home before dark, avoiding my mother's call out the window and the sincere apology she would have demanded of us for being late. We were minding our manners, wary of attracting her notice. (She was not averse to whacking her children with the back or her hand, or locking us into the pantry for misbehaving.) We were not the focus of her attention that night, however; my father was. Between bites of lard and tomatoes, I stole glances at him. His face burned and glistened with sweat. His eyes were glazed. And he wasn't talking. In place of the usual banter, all I heard was the staccato sound of the grandfather clock in the next room.

Up to the time Dr. Ganzing arrived after dinner, my world still ran on familiar certainties. Dr. Ganzing was a known quantity; my father rented space for his dental practice in a building the doctor owned in the town center near the *Marketplatz*. But this night the doctor carried his black leather case with him when he disappeared into my parents' bedroom with my mother.

The house ticked like a clock. I curled up on my bed, turning the pages of a book in the bedroom I shared with my brothers, Gunther and Rüdiger, but my ears strained for any noise from beyond the door. After a short while, I heard the door to my parents' bedroom open and close, followed by steps in the hallway. There was a murmur of voices, then the front door clicked and groaned as my mother saw the doctor out. The sound of her footsteps grew louder as she came back towards the bedrooms. Then the sound stopped. One of my brothers opened the door to our room. My mother was standing outside her bedroom door across the hallway. Then, she was not standing. She lay sprawled pale and motionless on the floor.

The three of us ran to her. "Mutti, Mutti" we cried, "Mom, Mom!"

After a moment or two, she got up. Steadying herself against the wall, she leveled her eyes at each of us.

"Vati will have to go away," she said. "To the hospital. He has something wrong with his brain."

My father was transferred to a hospital in Göttingen. Even though it wasn't a malignant brain tumor, as the doctors had first suspected, he remained in the hospital rehab unit for two months. Even then, he was too sick to return to us. He went to stay with his brother, my Uncle Rudy, for further rehabilitation. When he finally returned in early summer, the robust Vati I had known was gone forever.

It would be twenty years before I understood what had taken place in my father's brain in the weeks preceding that October night. What triggered the abscess was an upper respiratory infection, from which the inflamed bacterial or fungal material gravitated to his temporal-parietal lobe. There it festered and fattened into a vile pus-filled swelling. My father was unfortunate in that such an abscess was, and is, a rare condition. He was fortunate in that fifty years earlier it would likely have killed him.

By the time my father suffered what was, in fact, a seizure that October night in 1952, removal of such abscesses had evolved to a fairly sophisticated level. Ten years earlier the use of sulfonamides and penicillin had been introduced, and the surgical approach had advanced from aspiration and drainage to the "enucleation" or total extirpation of the lesion as though it were a brain tumor. As a white paper I came across from the early 1940s attests, "The devastating associated meningitis, suppurative cerebritis, rupture into the ventricle, which were the bugbears of all surgeons dealing with these legions in the preantibiotic era, no longer beset the neuro-surgeons."

Having said that, having a good section of your skull plate removed, your dura incised and peeled back, and an infected mass dislodged from your surrounding brain tissue is an ordeal. When my father finally returned home, he had clearly been run through the wringer. His face was thinner, and he moved like a boxer who had taken one too many punches.

I remember running to him the day he came home, expecting the usual robust greeting—"My little *Dicker* (fat one)"—and waiting for him to swing me up in his arms, but when I hugged him and took his hand, he stood motionless, like a man in a trance.

As the weeks passed, I had to accept that my longing for him to be the man he had been before his illness would not restore him. I had seen, for the first time, the vulnerability of the human brain and the ability of disease to bring a person down, no matter how strong that person had been before.

CHAPTER 4

A GERMAN-BORN DOCTOR

In every conceivable manner, the family is link to our past,
bridge to our future.
ALEX HALEY

TWO WEEKS FOLLOWING the surgery, TJ held steady. He was off the respirator and breathing on his own. The swelling had subsided and the bruises and scrapes on his face were healing. Although he remained comatose and encased in a complete leg cast and the rigid halo brace, his vital signs were good. All the staff had heard of the boy, and a feeling of expectation pervaded the entire Pediatrics ICU department, where TJ's bed was positioned out on the floor so he could be constantly watched.

Assigned to the pediatric service, resident Brian Fitzpatrick registered the first sign of purposeful movement. TJ waved his hand toward the place where the young doctor had pinched his chest—on both sides. Whereas TJ had moved on both sides as early as ten hours post-op, now he was not simply jerking an arm or leg in response to stimulus; he was reaching towards the place it originated.

Not long after that advance, TJ opened his eyes and looked around. We couldn't be sure he was seeing—that his brain was processing the images produced by the light his eyes were registering—but because both eyes were moving together, we believed that his brain stem had come back completely and that the connections to the cerebellum had been restored.

It wasn't much to pin our hopes on, but on May 16th, I stood in my blue scrubs behind a podium with a white-coated Robert Spetzler to my left and pediatric neurosurgeon Hal Rekate in suit and tie to my right. Reporters from the principal U.S. wire services, leading dailies, and local network outlets pressed forward, jostling with hacks from the supermarket tabloids.

It should have been a heady experience getting such attention from the press. I'd been featured in a local story five years earlier after operating on a young man who'd come into the emergency room with a 3 ¼ inch nail embedded in his skull. The twenty-year-old had bumped into a nail gun on the job, and hadn't even realized the nail was there until hours later. But the magnitude of interest in the TJ case worried me. Though I was impressed with TJ's progress, I felt uncomfortable. I've never liked counting eggs before they're laid.

As director, Robert fielded most of the questions that day, though Hal probably described the accident and the dangers it had presented, and I explained the procedure I'd devised to reattach TJ's skull to his spine. Robert is a superb surgeon and leader, and an eloquent champion of the Institute. I am not sure, however, that I would've displayed as much confidence as he did that day—at least publicly—when it came to our assessment of TJ's outcome and outlook. Robert is formidably optimistic. He's also a bit of a showman, as he soon demonstrated. Coming to the end of his statement he dropped a line that immediately reverberated throughout the media. "He's an absolute miracle child," Robert said. Hands shot up. The press barraged us with more questions. "What now? What's in store for TJ?" Robert didn't balk. He didn't hedge. "Timmy Mathias will not only recover all his neurological functions completely," he said, "he will walk out of St. Joseph's hospital. I don't know when but he will walk out of here."

When the event broke up, Hal and I reconnoitered outside the auditorium.

"What do you think?" Hal said, tapping the tips of his spread-eagled fingers together.

"Well, it might be a little premature to tell people TJ will walk out of here. He's still comatose. Still in the halo. And he hasn't spoken."

"Right," Hal said. "What is he now? An 8 out of 15 at best on the Glasgow Coma Scale?"

I nodded. The GCS is the tool used to ascertain neurological functioning. TJ was only halfway home. And even if his progress satisfied us doctors and gave us reason for hope, we all knew—as the press did not—the very uncertain nature of neurological recovery.

Still, what Robert had so boldly asserted is exactly what came to pass. At the end of May a nurse was tending to TJ when she heard a whisper. "Please help me," he said. They were his first intelligible words since the accident. In early June, a male nurse observed TJ reach with his toes for a teddy bear the nurse had placed at the foot of the boy's bed, and pull it up to where he could grab it with his hand. A week later, propped up in a wheel chair, TJ felt for the same teddy bear, now in his lap. He grabbed it and flung it ten feet down the ward. And on June 23, two months after his brush with death, TJ walked out of the hospital with the aid of a walker and went home. He'd progressed faster than even Robert could have predicted.

Hal and I viewed the most recent X-rays the day TJ checked out. As I'd banked on, the bone marrow from his hip was growing, helping fuse his neck to the occiput bone of his skull.

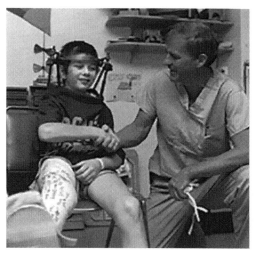

With TJ just weeks after the surgery. The press dubbed him "the miracle boy." Photograph used with the permission of Barrow Neurological Institute.

"It's remarkable," I said to Hal. "If someone had told me that night in surgery that this boy would not only walk out of here, but would do so a month after Robert said he would, I'd have said they were nuts. Completely nuts."

"And I would have been the last person to contradict you," Hal said.

We walked out to the lobby. "The whole thing is nuts," I said. Dozens of journalists had gathered for TJ's send off. They called out to him and snapped endless photos as the boy raced up and down for them in his wheelchair and demonstrated his skill with the walker. Under the metal band across his forehead—part of his halo brace—TJ smiled broadly. It was his hour and he was enjoying the celebrity to which he had become accustomed over the last month.

I'd become accustomed to celebrity too. Since the press conference my phone had been jangling with interview requests from reporters around the world, all wanting to hear how I had saved the "miracle boy." Sensational headlines had appeared, including the graphically inventive, "Miracle Surgery Saves Boy Who Had Head 'Cut Off' in Truck Accident" and equally compelling references to doctors saving the boy's head and the "nearly decapitated boy recovering."

But the reality that this operation was going to change my life didn't really hit me until I arrived home one day to find news vans parked in the street and reporters swarming about our home. Photographers scurried around snapping shots of my family and me. Even our dog got her moment of fame.

Aside from the wow factor in the surgery itself, the acclaim that followed in its wake brought with it additional recognition in the form of a nod to the country that had produced me. It was a "German-born" doctor, the articles noted, who had "sewn a boy's head back on." The German press made the most of that fact. The story about *Das Wunder* appeared in multiple publications, including *Bild*, *Stern*, *Bunte*, *Das Neue*, and a woman's magazine, *Echo der Frau*. Each story highlighted my German roots and featured photographs of me attending to the boy, examining X-rays, and explaining with a model how the surgery had been accomplished.

Among my family members in Germany, my mother's youngest sister was the most touched by the event. Tante Rulli had held me at my baptism in Prussia, as the Russians bombed craters around us at the end of WWII. Now she sent me the German publications in which the articles about *Das Wunder* had appeared. She told me how happy she was that she had lived to share in my success, and how proud my mother would have been to see it. I was a credit to my maternal side, the Albrecht family, she never failed to remind me.

My aunt's words provoked both happiness and a stab of loss and pain. My mother had died three years earlier. She'd lived to see me achieve stability in my profession after decades of struggle, and for that I was grateful. But I'm not sure she knew how much I attributed my success in life to her example of fortitude and hope. She had held the family together during my earliest years in a refugee camp, through my father's illness and our subsequent immigration. And now this formidable woman, to whom I owed so much for my success, was not here to share in it.

The newspaper articles brought to mind another story that appeared in the Arizona Republic a month after my family's arrival in Phoenix in 1957. In the staged photograph accompanying that article, my parents, two brothers and I are grouped around the kitchen table. My mother is holding some canned goods and "proclaiming wonders of American foods." My older brother Gunther holds a book to indicate how "puzzled over the language" we were. My father is quoted in the caption as "going slowly weak in the knees" and "worried over work."

The pleasant look on my father's face belies the creeping depression and confusion that had come with the residual effects of the brain abscess. Immigrating to a country where he couldn't understand the language only exacerbated his problems. My own bright smile reflects my innocence and ignorance of what it meant to be a German in America at that time. Soon after the photo was taken, I heard the word "Nazi" for the first time, when a neighborhood kid taunted me with the label. I had no idea what a Nazi was.

At the age of twelve I was too young to understand just how grim our situation was. I wasn't even sure why we had come to America. Retching from seasickness on the refitted troop transporter that had carried us across the stormy Atlantic, I had only a vague understanding of the events in my earliest years, and how they had led to our leaving everything behind.

I opened my eyes for the first time at the tail end of World War II, on November 23, 1944 in the old walled city of Graudenz, which was then in northeast Germany and is now the Polish town of Grudziądz. It was a city that was fast becoming a landscape of bombed-out craters and smoking ruins. Since the summer before, terrified ethnic Germans living in the eastern territories had either begun to evacuate or had been expelled, and by the time I took my first breath, thousands of ragged refugees were pushing their overladen carts through the snow, desperate to stay ahead of the advancing Red Army.

At the moment she pushed me into the world with the help of her sister (Tante Rulli) and her mother (my grandmother, Oma), my mother must have been frantically reviewing her next move. In the bitter cold of January 1945, she joined *Der Flucht*, as the flight of refugees was called, carrying me and pulling my five-year-old brother along. My father was attached to a military unit somewhere in the East, and she had not heard from him in weeks. Our survival was up to her.

Capture by the enemy was imminent, but at least there was a choice between two evils, and my mother must have known that it was far better to get rounded up by the Allies than the Bolsheviks. Like all Germans, I'm sure she had heard about the terrible reprisals at the hands of the Russians for the atrocities the Nazis had committed, and the reports of rapes and executions being perpetrated on Germans caught in the path of the Red Army. She managed to board a train heading west, juggling my brother Gunther and me and our few belongings.

I pieced together much of what I know about this period from stories my aunts told me or snippets of conversation or pictures in the photo albums that I would not have the heart to look at for decades. But history has also filled in many blanks. When I later read of how thousands of desperate people fled at the last moment because Hitler had forbidden the evacuation of eastern territories up to the end; of how thirteen thousand Volkdeutsche found themselves stranded as the lines shifted, while all along the Eastern Front, the Soviets were dragging those suspected of being Nazis out into the streets and shooting them on the spot; of the way Soviet aircraft strafed columns of terrorized refugees and blasted whole towns into rubble, I think of my mother and wonder how she survived.

Thankfully, she had the sense and drive to get us out while she still could. In early March, the troops of the 2nd White Russian Front took Graudenz. Five thousand citizens surrendered. In April, Hitler put a bullet in his brain. In May, half a year after my birth, Germany lay silent and filled with the stench of death. The country from which I had sprung was devastated. *Stunde Null*—"Zero Hour"—they called it. And *Stunde Null* it was for us, as for all Germans.

My mother with, from left, Gunther, me, and Rüdiger in a photo my father took in Bad Hersfeld. Our family was intact until my father's brain abscess put a halt to our personal postwar recovery.

When Graudenz fell to the Russians, it spelled the end of my family's time in the East. My mother, brother, and I ended up in an allied refugee camp, Lockstedter Lager, in northern Germany. My parents must have reunited briefly, because my younger brother Rüdiger was born there in March of 1946, but the few pictures taken of us over the next year or so show only my mother present. In her handwriting on the back of the photos are the words, "*Ohne Vati*, without Dad."

Only vague memories of that place surface now: pathways paved with white and yellow pebbles; a window; a rooster, my father with a shovel.

It must have been a year or so after we arrived because my little brother Rüdiger was already able to sit up by himself and my father had rejoined us. I was watching my brother through a window as he played outside. The next image, a vivid one, is of Rüdiger under a large rooster. The creature was furiously pecking at my brother's body and face. Then I saw my father running over with a shovel, and suddenly my brother was free.

Another memory is more pleasant. One day, my parents found a way to get away from the camp, which was near Hamburg. They took me to a park on the outskirts of the city that had a lake. Each took one of my hands and together swung me high and low.

A rooster. A lake. Of the four years we spent there, that is all I remember.

During our time in the refugee camp, my father was unemployed. Finally, after a long search, he learned that a city to the South, Bad Hersfeld, needed a dentist. On October 22, 1948, he moved our family there.

Far from the spartan barracks we had known, our home was now the entire second floor of a three-story, white, stucco box of a house. We had our own kitchen and bathroom, two separate bedrooms, and even an extra room presided over by a grandfather clock, some bookshelves, and a stern portrait of my father's father in his steel blue Prussian officer's uniform.

My earliest memories from that period are of food:

I was at the butcher's with my mother, gaping at the pink array of *Fleisch* and *Wurst* in the glass cases. I was hungry, which was not unusual, but I did not expect to get any of the meat in the case. My mother was probably at the shop to buy more of our main staple—*Schmaltz*, or lard. Suddenly, like the finger of God, the butcher's red pudgy hand reached down over the glass, dangling a piece of *Fleisch* in front of me. I grabbed at it before it disappeared.

Perhaps a year later, I had grown old enough to run down the street to the milk wagon, tell the owner how much we needed, and hold a pot under the spout while he turned the spigot. Carrying the pot back to the house, I was careful not to spill a drop. A glass of milk was like meat—a rarity, as was cheese, which we bought from the milkman too, but only once a week.

The cheese memory is fleeting and sensual. It has a soundtrack. The cheese was Camembert, nothing special in those days, and I wrinkled my nose and tried to blow the smell away through my nostrils. My father laughed. "The worse the smell, the better the cheese," he said.

And a last memory, improved by a photograph, is of my first day at the *Linggschule*. I had come down the path through the fields in my school-boy's uniform of short woolen pants, knee-high hose, and a button-down shirt. I was carrying home an oversized, colorfully decorated cone of sweets called a *Zuckertüte* that, along with the other new students, I had received that morning, and I was grinning the grin of one visited by un-believable luck. I paused to have my picture taken. Even now, when I see that photograph, I can taste the sharp sweetness on my tongue, and feel the supremely satisfying crunch of the sugar between my teeth.

The image of my father that I most like to remember is the vital, confi-dent man in his prime who smiles from the black and white snapshots taken around that time. That was the Vati (Dad) who was always jump-ing around with his box of a camera, recording birthdays and holidays and vacations. That was the Vati who presided over the lively Saturday night parties in the room off the kitchen reserved for special occasions, leading heated discussions while my mother sat on the sidelines smok-ing her *Haus Bergman* cigarettes, her *HB*s. That was the Vati who always had a home improvement project going, who loved to kick a ball around with his sons, who called me *Dicker*, the fat one.

That Vati had regained control of his world and had everything to look forward to. He had at last reclaimed his place in medicine, along-side his brother and father.

His father, Arthur Albert Sonntag, must have been an imposing man. I never knew him, but he is the severe figure in the Prussian uniform whose portrait held such a place of prominence in our home. One telling story about him—and about my father—has been passed down in family lore.

My father, Heinz Arthur Sonntag, was born in 1903. He was in his late twenties when he decided to become a dentist, and by then, my grandfather reportedly doubted my father's capacity to follow in his professional footsteps. My father had had a rather freewheeling youth; in his early twenties, he'd supervised a brickyard in Tanzania (then, since the defeat of Germany in World War I under British rule and known as Tanganyika), become a fisherman in Norway, and even driven his Adler to second place in the annual Nürnberg race. But by the mid 1930s he was ready to settle down, and he decided to apply to dental school.

My grandfather knew the demands of a career in medicine. An ear, nose, and throat doctor in Berlin, he seems to have gained some prominence and was said to have once seen the Kaiser in consultation. As the story goes, when my father revealed his plan to his father, my grandfather set him down for a reality check:

"You have been around the world," he said. "You went to Africa; you went to Norway; you have been a playboy; you will never make it through dental school; you will not have the discipline."

So sure was my grandfather that my father did not have what it takes to become a dentist that he made a bet with him:

"If you make it through dental school, I will drive down Friedrich Strasse, get out of my car, bow down, and take my hat off to you in the middle of the street."

When the day came in 1939 that my father did receive his diploma, my grandfather made good on his promise. I imagine the old gentleman stepping foot out of one of those grand, powerful touring sedans of the 1930s and, with an elegantly gloved hand, proudly doffing his top hat to his son in full view of his fellow Berliners.

My father's brother, Uncle Rudolph, was also an Ear-Nose-and Throat doctor. Unlike my father, he managed to survive the war with his finances intact. My father's only sibling, he'd had polio as a child and is said to have paled in comparison to my father in the early days—the flamboyant traveler who disregarded the conventions of his bourgeois upbringing. Still, in the postwar years, my uncle's more conservative approach to life had paid off. While our family was reduced to living in a rented apartment, Uncle Rudy owned a house and a car, and could afford to stay in hotels and dine in restaurants.

Uncle Rudy was reportedly the only person on either side of my family to have joined the Nazi Party. The story goes that he disengaged from the party early on, and we only found out from one of Uncle Rudy's sons, years later, that he had done so. I don't suppose that contributed to his success, but he seems to have had the knack to make the right moves in those treacherous times. One decision of his that proved to be a wise one was to have stayed in the western section, thus avoiding the turbulence and devastation that followed in the wake of the Russians.

Now that I had made headlines in the land of my birth, I thought of my father, uncle and grandfather. I was happy to have followed their footsteps into medicine. Happy to have made some Germans proud. But I was not a German. I was an American. (A distinction Uncle Rudy and his sons inadvertently helped me discover on my first trip back to Germany in 1965.) And if America was not a place, as my mother had assured me before leaving Germany, where gold lined the streets, it did turn out to be the "land of opportunity" that she had hoped to find, if not for herself then for my brothers and me—her "men." Basking in the glow of the TJ success, I wished fervently that she had lived to see me make my mark.

I also wished my father had survived long enough to see me achieve what had eluded him after the brain abscess took away his dexterity and peripheral vision. He had survived in this new country but had been deprived of the opportunity to reclaim his career as a dentist and a breadwinner.

I could never have articulated it in my youth, but now I believe that an unconscious desire to redeem my parents' sacrifices motivated me to succeed. That aside from the drive I felt to do something good with my life—to heal people and relieve suffering—my determination also had something to do with their example of tenacity and perseverance. When I reached success at last, a small hollow place in me rang with their absence.

CHAPTER 5

AN ATTENDING IN STEEL TOWN

Be it jewel or toy, not the prize gives the joy, but the striving to win the prize.
EDWARD G. BULWER LYTTON

WHEN I RECONNECTED TJ's skull to his cervical vertebra, I was moving away from the cranial and into the spine—in particular, instrumentation of the spine. But like all neurosurgeons, I'd cut my teeth on brain cases and rudimentary spine procedures during my residency, and had notched up over three hundred cases my first year as an attending. That inaugural experience had been ten years earlier in Youngstown Ohio, at a 900-bed hospital called Saint Elizabeth's.

Ohio was the place I put into practice all I had learned during my residency under Bennet Stein at Tufts New England Medical Center: digging out tumors from the brain and brain stem; picking out bloated wormlike tangles of blood vessels; wrestling spinal cord tumors from the surrounding fibers; digging bullets out of brains; and trying not to cry over all the children with malignant tumors.

They were all tough cases, but the children were in a painful category all their own. Most of those were primary tumors that originated in the brain itself, and those primary tumors, including glioblastomas that arise from the supporting cells of the brain called the glial cells, are not only the most common brain tumors but also the most malignant. I felt sick when I had a child come in with a primary tumor because even though children are more resilient than adults, if they had one of these, I knew they wouldn't survive long.

Along with the surgeries and lecturing the residents, Youngstown was the place where I also adapted to the enormous responsibility of calling all the shots for the first time. Even as a chief resident, in the back of my mind I knew I could call Stein or one of the attendings. Now that safety blanket was gone.

Two representative complex cases come to mind from that time: a young man with a sphenoid meningioma (a tumor at the bottom of the skull) and another with a cranio pharingioma above the pituitary gland. Both had rough recoveries, and one came down with diabetes insipidus.

But they both pulled through. One case, though, outshadows all the others that year for the sheer tragedy of it.

I had already driven home the night of the incident, through the gutted downtown around the hospital—it looked gloomier than ever in the deep dusk of late fall—and then on into the suburbs where Lynne and I had planted ourselves. I was eating dinner, but since I was on call, it was hardly a surprise when the beeper went off.

"What have we got?" I said, when I got the ER doctor on the phone.

"You need to get down here," he said. "A head trauma came in. A little boy."

I rushed back to the hospital, grim questions forming in my mind. Head trauma could be any number of things. Grisly scenes from my internship and residency went through my mind: the young man who had jammed a gun under his chin and managed to blow his frontal lobe off without killing himself. Another gunshot victim, this time a young woman, who walked in on her own steam while I was at Carny Hospital in Boston, blood streaming from the gaping, mangled cavity in her forehead.

As the suburbs gave way to the boarded up storefronts and empty lots around St. E's, an even more disturbing thought twanged at me. I had been an attending for just about three months. Intellectually, it registered that I was the one making the final life and death decisions and answering for the outcomes on extremely difficult surgeries. But so far the cases had been, if not easy, fairly textbook. Now the realization hit me again: the safety blanket of having a senior physician as backup was gone. I was the senior physician. The buck stopped with me.

My first thought when I saw the boy was, *He's dead.* He was a fair-haired child, three or four years old. His eyes were closed. His breath was so shallow as to be imperceptible. The top front part of his skull was shattered, exposing a stew of pinkish-grey brain tissue, bone fragments, and tatters of skin swimming in puddles of blood.

I swallowed hard. "What happened?" I said to the ER doctor.

He gave a snort, shook his head, and inhaled deeply. "Parents said he was riding in the back seat of the car. They were passing under an overpass when some kids above threw a rock down. It shot through the front windshield and right into the boy's skull."

"That's awful," I said, feeling the futility of the word to express the regret I felt. This child, who should have been at home sleeping in his own bed, lay here instead, near death. And why? Because of one careless, foolish act.

I knew there was no miracle that could restore the boy. But I was already pulling on my gloves to check for signs of brain activity. I placed my finger on one of his eyelids, pulled it up, and shone a light into his pupil. The full dark moon of the pupil did not constrict. The second eye yielded the same result. Next was the gag reflex. Ordinarily I'd have used a tongue plate to hold down his tongue while pressing another one down his throat. But the ER doctor had already intubated the boy through his mouth. I yanked gently on the tube, once, twice. Nothing. It was all but certain that the brain stem had shut down.

I turned to the ER doctor. "I'm going to talk to the parents. Then I'll need a general surgeon or a resident to assist. I'll call the OR and have them get the room ready."

I found the young couple in the waiting lounge and brought them into a private room. Hardly any details of their appearance come to mind now, only the hard basics. They were young. They were in shock. They were devastated.

"It looks bad," I told them. "We're going to debride the brain—to remove the bits of bone and skin. Then I'll stitch it up. I'll return afterwards. You can wait in here."

There was nothing more to say.

I spent the next two and a half hours trying to reverse the damage in the ragged crater that had been the boy's forehead and frontal lobe. With a thin, long-jawed bayonet forceps, I picked out the tatters of skin, the mangled tissue, and shards of skull. Normally I would have worried about infection from the debris in the wound. I would have worried about damaging the brain further while I dissected it trying to get every bit and sliver out. But now, as I scraped clean the intact bone, such concerns only magnified the futility of my efforts. When I could do no more, I set to closing the various layers: suturing the shredded dura and peracranium, and fitting back together as much of the skull as I could. It was like building a model with an incomplete set of broken pieces. Finally, I pieced his scalp back together, sutured it and stepped back to let the nurses wheel him off to the ICU. The awareness of my inability to save the child pooled in my mind, and began its slow, acid drip.

I returned to the parents. "It doesn't look good," I said. "There's a lot of damage to the brain. I'm very sorry. He'll be in the ICU. You can see him shortly."

And then I went home.

The boy passed away in the night. It was left to the ER doctor to deliver the brutal truth to the parents. At least I was spared the sight of the young couple trying to wrap their heads around the fact that their little boy was really gone.

Youngstown turned out to be the path, in a roundabout way, back to the academic career I had rejected at the end of my residency under Bennet Stein at Tufts New England Medical Center. In hindsight, that I even made the decision to pass up on Stein's offer is ironic. Even when he was still at Tufts, Ben Stein was a giant in the field—he later returned to Columbia University to head the Presbyterian Neurological Institute. And I think he recognized before I did that my future lay in academics, not in private practice.

I remember the turning point well. Lynne and I had, on a couple of occasions, been invited to Stein's home for dinner. It was on one of these visits during my last year in residency that he offered me the opportunity to stay on in the neurosurgery department at a teaching hospital connected to Tufts, also called St. Elizabeth's. I was flattered that Stein had offered me the position, but while I was talking to him about it, Lynne was in another room getting the low-down from Stein's wife, Doreen, and hearing a whole different perspective on what it would mean to us as a couple.

I brought it up on the way home in the car:

"Stein offered me the opportunity to take the position at Saint Elizabeth's," I said.

"Yes, I know," Lynne said, "but I was talking to Doreen. She told me not to let you go into academics. She said that if I thought I didn't see you much now, I would see you even less if you did that."

"Well, that might be true for the first five or ten years, until I get established," I got in before Lynne stopped me.

"No, Doreen said that if I thought there would be a light at the end of the tunnel, I could forget it. She said that when you're married to a neurosurgeon, especially in academics, there is no light at the end of the tunnel."

So, after that night, I listened to Lynne and started looking for opportunities for private practice with a neurosurgical group or as an associate. A search like that takes you all over the map, and it's a bit of a crap shoot finding an appropriate opening that is also in a city where you want to live.

Naturally, one of those plum spots showed up early, an opening in San Francisco. I applied for it and lined up some other possibilities: one in Cooperstown in upstate New York; a doctor in Waco Texas who was looking for an associate; and another in Corpus Christi. Finally there were two general neurosurgical practitioners who had formed a group in Youngstown, Ohio, Doctor Kalfas and Doctor Kholi.

I did get an interview in San Francisco and was hopeful about it, but I wouldn't know for a while. As for Corpus Christi and Waco, I don't know if it was the provincial nature of the cities, or the fact that they were hot and humid and prone to hurricanes on the one hand, and tornados on the other, but we crossed both of those cities off the list without too many misgivings. That left Youngstown, Ohio, a city that, in 1977, still called itself Steel City, USA.

The first interview in Youngstown went well, so I brought Lynne with me to the second interview, which is not uncommon, since—as my long experience has borne out—a doctor's spouse is going to have to support what you do or it's just not going to work. Either your marriage or your career will suffer if you don't make these big decisions together. Both Lynne and I were happy enough with the quality of the group, and we were particularly impressed with Dr. Kalfas, a kind of Renaissance man of Greek origin who spoke several different languages. Still, right from the get-go, Lynne did not want to go to Youngstown.

It was a dismal town, maybe still is. It recently made *Forbes' Magazine*'s list of the twenty most miserable cities in America, along with Detroit, Cleveland, Toledo, and Flint Michigan, among others. Like those industrial centers, it was characterized by gutted factories, air that stank of hot steel, high crime, unemployment, and lousy weather—hot and humid in summer with bugs like a Biblical plague, and bitingly cold in winter.

After keeping us waiting for weeks, San Francisco, our first pick, ended up not filling the neurosurgery position that year, and Kalfas and

Kholi made me an offer that was too attractive to turn down, especially in view of the lack of other opportunities. Lynne was in Phoenix visiting her family when the choice became clear to me. As I dialed the phone I could feel the news I was about to break sticking in my throat. Lynne tried to sound upbeat, but her apprehension crackled over the wire.

"So, how did the third interview go?" she said.

"Well, they made me an offer it'll be hard to turn down."

"Go on."

"Fifty-five thousand a year and a new car, and I'll be an attending neurosurgeon at a new medical school being built in the nearby town of Rootsville. It's a good position Lynne. We're just starting out."

There was a brief silence. Then, "Oh please, not Youngstown," she said. "I'll go anywhere else."

But Lynne and I both knew there was no "anywhere else." We accepted the offer and landed in Youngstown that July. It was just before what the locals still call "Black Monday," the closure of one of the large steel mills and the event that started the whole steel industry on the fatal downslide that would lead, decades later, to a place on that distinguished Forbes' list.

Youngstown cemented my experience in taking full responsibility for the outcomes on tough cases, but it proved to be short-lived because of two distinct challenges. One had to do with Lynne.

Lynne and I had been married for four years. I'd fallen hard for her during my internship year in Tucson when, as a nursing student, she'd come into the hospital room where I was doing rounds with some med students. Still, three months later I'd driven away, leaving her in front of her parents' house crying in the rain. I was headed to Boston for my residency, and I felt I had nothing but uncertainty to offer her. Two years later on a trip back to Phoenix, though, we ran into each other again at a party. Within a week we were married, and she came back with me to Boston to rough out the intense remaining three years of my residency with me.

She'd been on her own in Boston much of the time. Now in Ohio, with my long hours, being on call every third night and then coming back in when an emergency hit, she was on her own again. She had felt isolated in Boston, but here it was worse. Lynne had gone back to school to become a nurse practitioner in pediatrics, but few people in Ohio even knew what a nurse practitioner was, and the doctors that did were extremely resistant to the idea of giving up any of their patients (or the income) to someone in this new kind of position. It might work in liberal Boston, but not in the Midwest. "They seem to consider the Boston model witchcraft or something," Lynne lamented.

When Lynne did finally land a job, it was at a small rural clinic seventy miles north of Youngstown on the shores of Lake Erie. We were expecting our first child by then, winter was setting in, and she was commuting all that way through snow and ice. It rattled Lynne and worried the socks off me.

We both liked the suburb we lived in, Boardman, which billed itself as the "green oasis of the community," and there were other pleasant towns in the metropolitan area, but Youngstown itself was an urban ghost town. It seemed that every storefront in a five-mile radius of the hospital was boarded up. Saint Elizabeth's, a huge, sprawling institution, was just about the only thing breathing downtown at all.

Over time, a much more serious concern emerged. I had still not taken my oral boards, so every Wednesday I was driving into Pittsburgh and sitting in on the Grand Rounds at University of Pittsburgh Medical Center, which had a neurosurgery department. I grew increasingly aware that I was in a backwater when it came to neurosurgery. I was handling what cases did come in, but they were not cutting edge, and since Doctor Kalfas and Doctor Kholi were primarily interested in private practice, not academic neurosurgery, I wasn't getting the collegial support necessary for me to advance my understanding in my chosen field. When the doctors informed me they planned to split up and that I would have to buy in to the practice, I knew things had come to a head.

Lynne and I reviewed our situation: There was the isolation from family in a scrappy steel town; the extremes of the climate—it snowed on May 15th that year, and Lynne found the giant summer bugs and the constant grey sky oppressive; my feeling shipwrecked professionally; and the discovery that Lynne was pregnant. That last development clinched it. My father had died, but my mother was still alive, as were Lynne's parents. We both wanted to raise our children around family. We made the decision to return to Phoenix.

We just had no idea how we were going to pull it off.

CHAPTER 6

AN ITINERANT NEUROSURGEON

Patience and tenacity are worth more
than twice their weight of cleverness.
THOMAS HUXLEY

I HELD THE LETTER in both hands, looking at the return address: John Green, M.D., Barrow Neurological Institute, Saint Joseph's Hospital and Medical Center, Phoenix, Arizona. I had been expecting it. At least hoping for it. It was a response to a letter I had sent off once Lynne and I had made the decision to return to Arizona. The contents could make the difference between a clean reentry to a city I had left ten years earlier and a future of uncertainty and debt.

Dr. Green was the founding director of Barrow Neurological Institute. His achievement in forging a top-rate center for neurosurgery in what was then a wide, neglected swath of the American Southwest became his lasting legacy. He'd come to Arizona from Chicago after World War II, the first neurosurgeon in the state. He'd rattled over miles of dirt roads and flown deep into rural backwaters to handle head injuries throughout the 1950s, all the while nurturing a wild dream of building a neurosurgical institute in the burgeoning hub of Phoenix. In 1962, that dream materialized through the unexpected windfall of a donation from coal magnate, Charles Barrow.

I didn't know any of that background then. And I didn't know Dr. Green from China. I only knew he was the chairman at BNI. And a position at BNI in the late 1970s would have been a phenomenal break. I thought that my background with Stein and the three hundred cases I'd racked up in Youngstown gave me a shot at it.

Barrow was not the only door I knocked on, though. Before leaving Ohio I surveyed all my possible connections and started the mail campaign that included Green. Back in Arizona for Christmas in 1978, Lynne and I made a side trip down to Tucson where I met with the chief of neurosurgery, Phil Weinstein, at University of Arizona. I had gone to med school at U of A, had delivered the convocation for the very first graduating class in 1971. Nothing. A couple more trips back in the following months were just as unproductive as the first one.

And now the letter from Dr. Green. I stood at the table in the kitchen. Lynne joined me, her eyebrows raised in expectation, and hope. The shadows were lengthening beyond the windows that overlooked the sloping, wooded yard. I ripped open the envelope and scanned the letter quickly.

"Well, what does it say?" Lynne asked.

I gave a bitter snort. "It says, 'If you want to practice neurosurgery in Arizona, you will starve.'"

A few months later, I sat in my car in the sweltering heat of an Arizona August. I was heading to Desert Samaritan hospital on the east side of the Phoenix metro area. The four o'clock sun beat on the roof, turning the car into a convection oven. The leather seats felt like heating pads. As the baked cityscape passed by in a heat-heavy haze, the hoped-for outcomes of my plan drummed through my head: *I've got to get privileges at a hospital. If I can get privileges, a hospital can call me to take emergency trauma calls. Local doctors will get to know me and send other cases my way. And I'll be able to build my practice.*

We'd left Youngstown the June before—despite the gnawing question of a position at the other end—me setting off in a U-Haul through a driving rain, and Lynne flying back with our two-week-old baby daughter Alissa. I'd managed to break my leg the weekend before our departure playing soccer against a team of burly football players who only engaged in the sport to keep in shape during their off season. On moving day, I remember gimping out to the truck on my pole-of-a-leg, hoisting myself into the cab, and driving into a future that was about as clear as the sludge in Youngstown's Mahoning River.

The lines I had cast from Ohio had turned up zilch. The only response that left any wiggle room for hope had come from a ruddy, quiet guy I'd gone through med school with, Bill Reid. Bill had done his neurosurgical training in Dallas but had returned to Phoenix. "You can come out," he'd replied, "but I am not doing anything. I guess we can do nothing together." So, Lynne and I had cleared out of Ohio and settled in a rental house in northeast Phoenix, with a new baby, a mounting pile of bills, and a mortgage on a house in Ohio that we weren't even living in.

I had put into immediate effect my one fallback plan. Pushing back the thought of what I'd been bringing in at Saint Elizabeth's, I made a deal with the director of surgery at the County Hospital: For ten thousand dollars a year, I'd do every neurosurgery case that came in.

I'd also come up with an action plan. My first big challenge was networking. I couldn't sit around waiting for a call to strike like lightning. I knew how a hospital was run, so I'd decided to attend the monthly doctors' meetings at hospitals all over the Valley and get the word out that I was here. I would hand out my business cards and schmooze like a car salesman.

But now the kinks in my plan revealed themselves. The temperature had topped 100 before the morning was out. And our Audi, which Lynne's brother had driven out for us from Ohio, had no air conditioning. Beads of sweat broke and trickled down my temples and into my shirt collar. Beneath my suit coat, my shirt plastered itself to my back. I turned in at the hospital. As I walked across the parking lot, the heat bore down from above and deflected back up from the paving. My feet pressed into hot asphalt the consistency of a fresh-baked cookie.

Inside about a hundred doctors were taking their places at round tables waiting for the pre-dinner presentation to begin. I made my way over to a table of OBGYN guys—a woman might mention to her OBGYN that she has headaches, which can turn out to be the result of a pituitary tumor, leading to a referral to a neurosurgeon. I was trying any angle I could think of to get some referrals.

"Is this seat free?" I said. "Sure," the doctor to my right said, glancing at me and resuming his conversation with the guy on the other side. I pulled out a chair, conscious of the sweat cooling on my brow. My nervousness didn't help. I sat for a moment, letting my internal temperature regulate. Another doctor took the seat on my left, nodding at me as he pulled out the chair. He stuck his hand out and introduced himself.

"I'm Volker Sonntag," I said. "Neurosurgery. Pituitary tumors."

"Right," the doctor said.

A silent *So what?* echoed in my head. What did I expect anyway? *Hooray, we've got a neurosurgeon here!* I sat through the presentations and ate the generic dinner. I stuck around a little bit afterwards. But all I could think was the great impression I was making, if I was making any at all. *Sonntag? Oh yeah, that guy with the frizzy hair who was sweating like a horse.*

I did pick up one tip from those initial attempts to network, however. I immediately traded in the Audi for a Vega with air conditioning.

The Vega was an improvement but it did nothing to solve another problem, one that involved a second feature of my environment: distance. Greater Phoenix was, and is, a city of bedroom communities that have completely developed around the automobile. Compared to other major cities, there was no real downtown to speak of, and the suburbs were small cities in their own right that had grown up independently. Some of them, like Avondale in the West or Chandler in the East, were as far as forty miles out from where I lived. There was only one freeway that bypassed most of these communities, and taking surface streets required considerable travel time. I was dead in the water when I requested privileges at a Mesa hospital out in the East Valley. I got a similar response at Boswell, out on the West side in Sun City. "You're too far away," they said. "There's no way you could get here in time."

There were a number of hospitals closer in, so I kept making my rounds in the Vega, all the while feeling like I was in a vacuum. The phone never rang. When I came back from my wanderings during the day, radiated from the sun beating through the car windows and unable to suppress the hope that just one of the guys I had met would throw me a scrap, I was met with a deadening silence.

Returning home after one of those first disappointing outings, my mind turned, as it often did, to thoughts of my father that first summer in Arizona. He had met with closed doors too on his forays out into a world far stranger and more inhospitable that the one I faced.

We had arrived in "the Valley of the Sun" in early April of 1957, after a stormy ten-day trans-Atlantic crossing and three days journey overland by train. I'd spent most of the voyage pale and prostrate in the sickbay with an I.V. stuck in my arm. But at least the nausea was something that would disappear at our journey's end. The voyage over on the General W. C. Langfitt marked just the beginning of my father's new trials.

Unlike my mother, who had pressed for our immigration to America and welcomed the challenge with a sense of adventure, my father seemed to fill the empty shipboard days reflecting on how far he had fallen. The few photos taken on board the refitted troop transporter show a depressed, middle-aged man in suit and tie. In one, his hat is off, and he looks vulnerable with his balding head exposed to the brisk air. The identification tag he had been given upon boarding— made of stiff card stock, the kind used to advertise sale items—is clipped to his wide lapel.

One day I overheard some comments he made to my mother. We were up on deck, sheltering from the wind on one of the long, hard benches against the sweating, cold bulkhead.

"What have we come to Gila?" I heard him say. "Look at us. Crammed in with all these . . . refugees. My God, sleeping with them. How did we end up here?"

It was a question I think he posed to himself for the rest of his life.

A trim, upright American Lutheran named George Beuchel sponsored our trip. The husband of a German woman who had helped my mother in the house before immigrating to America herself, Beuchel owned a trailer park on the grounds of which stood a sole detached house. That was our new home. Like the houses we had seen on the ride over from the train station where George Beuchel had met us—along with a welcome committee of women from the Messiah Lutheran Church—the house was a modest, boxy building constructed of cement blocks painted in the pastel hues of a lemon butter mint. I'm sure the home in itself would have satisfied my parents. But the exotic landscape on which it sat gave it a strikingly exotic appeal. Strange flowering plants dotted the grounds: creeping lantanas, star-like

My parents Heinz and Gisela Sonntag aboard the General W.C. Langfitt, the refitted troop transporter that brought us to America in 1957.

daisies, and thick-stalked oleander bushes tall as trees. A large shade tree shielded the front of the house, and our own palm trees shot up like flagpoles on one side. Best of all, across the street was an orchard with rows of citrus trees stretching into the distance. (My astonishment at those oranges has become family legend. "Look!" I said, poking my brother Gunther. "Oranges! On trees!" We'd only seen oranges before at Christmas, a special treat in the center of the holiday plate of sweets we each received.) We had landed in Shangri-La.

But that was in April. By June we began to realize what it meant to live in the desert. Worse, the harsh reality that my father was unlikely to find work in anything resembling his profession as a dentist had sunk in. It wasn't only his lack of English. Though we didn't realize it at the time, the brain abscess had permanently impaired his peripheral vision and his ability to plan and coordinate movements. At the age of fifty-four, alienated from all he had known, my father found himself with no other choice but to offer up the only thing he had that could be turned into cash: his labor.

Apart from our house—the only permanent dwelling in George Buechel's trailer court—only a couple of cabanas had been constructed on lots where the trailers could be parked, bare-bones dwellings with a bathroom and living room. Mr. Buechel's plan was to populate that whole lot with cabanas, and he needed someone to help him do it.

Unfortunately, the dwindling of my father's hopes at finding a job suitable to his education coincided with the start of summer. It hit us like a blast from an open oven on high heat. With no leads from any corner, my father and older brother Gunther started digging ditches and hooking up pipes on the lot. I acted as "gofer," eventually graduating to helping the men position pipes in ditches. We labored like that all through that first summer. By the end of it, we had been relieved of our fantasies of a Shangri-La. And my father had resigned himself to his new life as a laborer and unskilled worker.

Things never improved for my father—and weren't likely to improve. The implications of that sad reality hit home for me two years later, the summer before I started high school. By then the best he'd been able to do was find work in a hardware store downtown, and even that wouldn't last. My mother had had a falling-out with George Beuchel's wife, and we'd been forced to move to a place we called "the Shed."

It was late in the season. My father and I had walked over to a public swimming pool and were sitting on a grassy slope, not paying much attention to the kids splashing around in the water. I wasn't paying much attention to him either. I was worrying about entering high school in a couple of weeks, and probably calculating how much money I would need to pay for supplies. Then he spoke:

"I never should have married your mother," he said.

His words cut through my thoughts with a startling brutality. I couldn't be sure I had heard him right.

"No," he sighed, "I never should have married her. I never should have let her talk me into coming to this place."

I looked at him. He was nearly bald. He was thin. His dark bushy eyebrows bristled over eyes hollow and dark with resignation. The scar on his temple burned white against his sunburnt skin.

"It was a mistake to marry her, *Dicker*. It was a mistake to come to America."

I didn't know what to say. I sat there with him, sweating in the heat, breathing in his disillusion with the hot air.

Now those memories assailed me. A feeling of failure and futility washed over me. I wondered if the dreams for my own future were destined to spool out in the same futile way. With no response to my inquiries, feeling anxious and frustrated, my mind slipped back onto the old familiar track of worrying about money. *How many times had I been to the Valley Bank since we got back? What is the debt up to now?* There was no way around that, though. Even though Lynne was able to find a position at Good Sam Hospital, her salary couldn't cover rent, food, and—of critical importance—the tools of my trade.

That was the big expense looming before me, my surgical instruments. At every hospital, a neurosurgeon had to have his own set. If he didn't, he couldn't operate. So, I went into deeper debt buying instruments for all the neurosurgical procedures I would be called on to do—dissectors, scissors, a Kerrison punch, aneurysm clips, spreaders, a Hudson drill. Only the knives came courtesy of the hospital. Then I bought a fishing tackle box to cart them around in, and made sure to have my initials etched into each instrument. I didn't want those puppies to wander away into anyone else's kit.

The waiting game went on for months. I kept doing the county cases; kept trying to build a practice. I was thankful to have Bill Reid as a partner, someone to split the business expenses with, even though it made my slow progress ever more palpable. When our time in one office coincided, we passed each other silently in the hall, nodding curtly at each other, and wishing the damn phone would ring.

One day in January 1979, Bill walked into my office. Bill was always calm and quiet, but on this day something else was going on. I looked up from my desk. His pursed lips and tight grimace told me he wasn't going to be delivering happy news.

"I don't think we're going to make it," he said.

"Oh?" I said. "OK."

"My former chairman connected me to someone in Tennessee. Knoxville. They offered me a job. I'm taking it."

I shrugged. "Well, good luck," I said. That was all there was to say. As lousy as I felt to hear the news, I couldn't blame him.

Within a week Bill was gone, leaving me with the lease on the two offices and a used Xerox machine that managed to either mangle or smudge the few forms and letters I was producing. In clearing out, he missed one last footnote to that whole sad period.

It was right around then that I did finally start seeing a few patients. My name was finally getting around. Surely I could start pocketing some profit after the bills had been paid. I hired a young woman for the Phoenix office to answer phones and keep records. As the months went by, though, I kept ending up in the red. *The bills on all the debt must still be outpacing my profits,* I thought. It took several more months before I finally put two and two together. I had been so used to being in debt that it didn't occur to me I was being robbed. That girl was cooking my books. In the end, she embezzled $17,000 from me. I had had my first lesson on the business end of practicing medicine.

I did eventually crack the nut and started getting cases through sheer persistence, through showing up at conferences, talking to other neurosurgeons about cases, and making myself available for consultations and emergencies. This led to privileges for ER calls at Good Sam. It went like dominoes after that. Privileges followed at Phoenix Memorial,

Scottsdale Memorial, and John C. Lincoln. Finally, I nabbed the big one. They didn't make it easy. I had to do fifteen supervised cases, all different types of surgeries, at Barrow Neurological Institute at Saint Joe's before they would give me unsupervised privileges. Once I had earned those privileges, I could take ER calls. I was a player at last, and I felt like I'd won the jackpot.

CHAPTER 7

THE ORAL BOARDS
AND OTHER ORDEALS

Seeing much, studying much, and suffering much
are the three pillars of learning.
BENJAMIN DISRAELI

IT WAS A CRISP, bright January morning. Lynne was pulling the car out while I grabbed my coat and briefcase. I had organized my study notes one last time the night before, then closed the door on the guest house in back where I'd holed up over the Christmas holidays. In the clear light of day, the confidence I had felt over the last week evaporated. *I couldn't possibly have prepared well enough*, I thought. *If only I had another week.*

It was January of 1981. The year before, I'd received notice that I was scheduled to take my oral board exams. I'd felt satisfied at clearing the first hurdle. I had the three hundred consecutive cases I needed, each written up and with a three-month follow-up on the individual patients; I had letters of recommendation from every hospital that I had practiced in, affirming my good standing; the letter from the program director at the New England Medical Center confirming that I had finished my residency in good standing; and proof that I had passed the written part of the boards. Now all that stood before me and my place among the ranks of the duly certified, was a last three-hour ordeal.

Lynne drove in silence for ten minutes. In the back our baby daughter Alissa emitted gurgles from her car seat. I brooded. It was a typically glorious winter day in the desert, but the bright sunshine only deepened my glum mood. Five minutes out from the airport, Lynne spoke.

"Well, the time has come," she said. "I'm sure you'll do fine, Volker."

"I suppose," I said. I glanced at her. Beautiful, practical, supportive Lynne. Always calm. Always so sure of me.

"It's natural to be nervous about the outcome and its effect on your career," she said.

"That's not it," I said. "I think my career can withstand my not passing the oral boards, but I don't know if I could ever look Dr. Stein in the eye again."

The orals are yet another of those excruciating benchmarks that test your ability to store and recapitulate knowledge. Exams like that also test your ability to withstand the kind of pressure that could give most people a stroke. And even though it is possible to practice without passing the board exams, if you want to practice at a reputable, big hospital, you need the boards; if you want to join organizations such as the American Association of Neurological Surgeons, you need the boards; if you want to teach residents, you should certainly be board certified. And if, somewhere down the line, you get sued, and the lawyers find out (which they will) that you are not board certified, they will have a heyday. *Doctor, I understand you are not board certified. And you are practicing neurosurgery? Why are you not board certified? Isn't it routine that you have to be board certified to measure up to the criteria of your profession?*

Today in the US, there are about 3800 board certified neurosurgeons and maybe another four or five hundred who are not. That last number is just a guess, because those neurosurgeons only infrequently show up at the association meetings and other events. They are off the radar.

Even back then, I certainly did not want to be counted in that ghost cadre of uncertified neurosurgeons. I put everything on hold for a few weeks. This was right before Christmas, and the exams were to be given in Memphis in mid January. I was building my caseload and had patients that needed to be seen, but I had an arrangement with a fellow neurosurgeon named Jack Kelly who agreed to cover for me.

Still, despite all the preparation, my potential embarrassment in front of my mentor wasn't the only thing gnawing at me. Tests always brought back a host of doubts. It seemed I had been taking—and sometimes failing—these tests for years. Each time I approached one, a reel of memories started up, accompanied by a nauseating apprehension and doubt.

One memory always started the movie in my head going.

I was picking my way through a dirt lot under the hot September sun. The baked soil beneath my feet was rutted with tire tracks from Caterpillar earthmovers and graders and strewn with plaster and rubble. Small mountains of pipes and lumber and dusty flats of plasterboard and bricks rose up like hurdles in an obstacle course. A short distance away, the iron skeleton of a new complex of buildings, girdled with scaffolding and catwalks, threw a sharp geometric pattern against the blue sky.

It was my first semester at U of A. The medical school was still being built and there was no hospital nearby where students could do their clinical work. The entire area was a construction site removed from the main campus. I had a warm baloney sandwich in my pocket and nowhere to eat it. The thick tie I was wearing, as I always did on a class day, was a noose around my neck. I had never been so discouraged.

The source of my angst was the "C" I had just received on the first test in my physiology class. I had never received a grade lower than a "B," and had mostly gotten "A's." Just a few months before, I had left ASU as the first recipient of the Fred G. Holmes Memorial scholarship. I had kept up a 3.60 GPA. The *Tempe Daily News* had done a story on me: "ASU Pre-Medicine Student Young Man with a Purpose." Now, here I was at the beginning of my first semester and I couldn't even pull a decent grade on a test. Everything I had thought about my education was being put to the test. Maybe I didn't have the smarts to get through med school.

If I had thought undergraduate work at ASU was a jump up in my intellectual understanding, U of A Medical School was another universe. It was also a rude awakening. It had only been my ability to tie myself down and study long hours that had allowed me to get the grades I needed to enter med school. Now I found myself surrounded by students who not only studied just as hard as I did but who were smarter to boot. The "C" was proof of that.

To make the odds of success higher, there were only thirty-two students in the incoming class of 1967 at U of A. All of them were top of their class in their undergraduate studies. The thirty-two of us shared nineteen professors. Considering the course load and grading curve, I was in for a long, hard haul. That's what I was thinking about as I crossed that lot.

That had been only one test however. The second more serious failure had come at the end of my second year in med school. We were still in the classroom most of that year. We took a class on histology, studying the different kinds of cells and their relation to one another, and another on infectious diseases. The most difficult class was pharmacology. I think I burned the entire *Gilman and Goodman Textbook of Pharmacology* into my brain.

The main thing I remember that year was a sense of impending disaster. Each class required you to memorize so much complex material that you felt your brain would explode. The second year was crucial. At the end of it, the USMLE—the medical licensing exam—was waiting to push those who couldn't cut it over the edge.

The USMLE is a critical exam that all med students must pass to get their medical license. Comprised of three parts, the first is given at the end of the second year, the second at the end of the fourth year, and the third after the internship year. I studied my brains out for that exam. I knew my future was riding on it. Still, when the day came, I failed the first part of the test.

It's not like that was the end of the line for me—you could take the first part again after your junior year—but failing it threw a real pall over everything. It made me question my direction every day.

Worse than the academic failures was the way the rigid mental discipline required of these tests seemed to summon a corresponding and opposite chaos. It was hard enough to absorb and retain knowledge of such complex nature. You also had to buck up and develop the fortitude to ride the external forces that constantly threatened to push you off track.

I'd had rich experience in this area. In the general category of automobiles alone, I could have veered off the track at any moment and not found my way back. When I was a boy, my family had had a string of unfortunate incidents with cars, but my first serious encounter came during a cardiology externship at Georgetown University in Washington DC the summer between my first and second years of med school—and that accident followed after a near miss of an even more violent nature.

It was 1968. I'd driven down to DC with two friends and fellow students from ASU—Steve Hall, who called himself the "Kingpin," and pipe-smoking Steve "Beetle" Bailey. This was right after Bobby Kennedy got shot, which had followed on the heels of the assassination of Martin Luther King Jr. Reverend King, along with the Southern Christian Leadership Movement, had organized a civil disobedience Poor People's Campaign designed to bring attention to the issues of poverty and hunger that had gotten overshadowed by the Viet Nam War. A key element of that campaign was the establishment of a tent city on the Mall in DC.

Once we got to the capitol, we knew we were out of our element. The climate was calm on the surface, but the tents were still up. Just the April before, the hospitals in the area had treated more than three hundred patients injured in the riots that erupted following Reverend King's assassination. The images of the looting and the bombed out cars and debris-covered streets stayed fresh in our minds. We made sure to wear our white coats wherever we went, figuring (hoping) that no one would give doctors a hard time.

One night about nine o'clock the latent violence caught up with us. Bailey, Hall and I had met up with a third guy from school, Chuck Rolle. We were sitting back, shooting the breeze at a pub called McGuire's in a gritty part of the city, when the sound of the door opening caught our attention. Immediately, the four or five other patrons in there, all working-class guys, jumped up and ran for the door. We didn't know what the hell what was happening. Then we saw the dirty bottle with a smoking fuse rolling into an open space among the tables. "Shit!" someone yelled. "It's a bomb!" We hightailed it out of there. A moment later, an explosion ripped through the place and smoke came billowing out.

That incident happened not long before we left the city. But the residual chaos still had us pegged.

Nothing much happened the first part of the trip back, once we navigated out of the congestion and hit the I-70 somewhere in Ohio or Indiana. The next day, the flat nothingness of Kansas and Eastern Colorado went by in a blur. Then we hit the foothills of the Rockies. The weather got bad. Bailey was driving and I was riding shotgun in the front seat when we reached the outskirts of Denver.

Rain was clattering on the roof, dashing buckets of water against the windshield and rendering the wipers about as useful as ribbons. We must have hit a slick, or maybe the wind caught us on a curve. Whatever happened, I felt the car starting to tip on my side. The next thing I knew, I was out in the rain on the ground, looking up at the car hovering above me on the two wheels that still had contact with the road. I had just enough time to think, *I'm a goner*, when the car stopped its trajectory, teetered in the air, and rammed back down the other way.

I don't remember what happened next. I expect Bailey thought he'd killed me. He and Hall jumped out of the car to find me. Once I realized I wasn't dead, it was all I could do to get my heart to stop pounding like a jackhammer. All I remember is the three of us getting back in that car and limping into Denver where we found a shop that could repair the damage.

It was the summer before my junior year, though, when I had an accident that nearly took my head off.

I had stayed in Tucson to do a neurology rotation with Dr. Bill Buchsbaum, the professor who influenced my decision to go into neurosurgery. Rotations are a kind of practicum that starts your third year. You connect with a doctor in a particular field, do the rounds with him, discuss the research, and then go back to the lab to conduct your experiments and record the results. I was getting a head start on it as an extra project for summer.

With my parents in Phoenix on a visit home from med school, ca. 1970.

I had driven over in my VW Bug to the Tucson Medical Center. I had done my rounds and discussed the project I was going to do. Now, on my way back to the medical school, I was thinking about how I was going to inject mice with steroids and wondering what results I would get. I was driving west on Elm and had entered the intersection at 6th Avenue when a vehicle coming south blasted through the stop sign and T-boned the Bug, ramming into the car on the driver's side. The impact catapulted me through the closed convertible top and clean out of the intersection. To add insult to the injury, I landed in a cactus garden on the northwest corner of the intersection.

The ambulance took me back where I'd come from, the Tucson Medical Center. I floated in and out of consciousness. At one point I recognized my good friend Sam Shoen's voice coming from a long distance. Another college and med school friend, Sam was then an intense, black-haired intellectual with mutton-chop sideburns and a fiery political consciousness; compared to me, he was a radical liberal, but our heated conversations only cemented our friendship and respect for one another.

They sutured me up in the emergency room. I was comatose during that procedure and for what followed. They wheeled me into the ICU, where I lay for a small eternity as the nurses in their starched white caps painstakingly picked the cactus spines out of my face and arms and torso and legs.

For two days I was in a strange, timeless, black hole from which I periodically surfaced before sinking down again. At one point, a lawyer named Feldman appeared. At another point, when my vision cleared, I realized I was looking at the scene of the accident on the evening news. The Bug was lying in the intersection, a heap of twisted metal and broken glass. On the third day I woke briefly to see my father and younger brother Rüdiger by my side.

When they let me out, I stayed at Sam's for a couple of weeks, waiting for the Frankenstein sutures on my head to heal. My back was a roadmap of abrasions and the cuts and scrapes all over my body had to be cleaned. But time was passing, and though my professors had been sympathetic, I knew I needed to get back on track with my studies.

I returned to my small apartment and got down to organizing my affairs. The hospital staff had given me a bag with my clothes and belongings. I hadn't thought about it while I was at Sam's, but once I got home I pulled all the stuff out. Funny, there was only one shoe. I went back to the hospital.

"I was in here a few weeks ago," I explained. "When I got home, I realized I only had one shoe. Did you find a shoe here that matches this one?"

They shook their heads, looked at my admission papers.

"You came in here with only one shoe," they said.

That was weird. Where the hell was my other shoe? Then it hit me. I went back to the intersection. There was the cactus garden I had been thrown into. I circled around and there was my shoe, impaled on the spiny arm of a cactus.

It had taken me weeks to recover from that last accident, and even longer to feel confident I was still operating with all my marbles. But, I had regained my stride. No matter how insurmountable the setbacks seemed, experience had taught me that hard work and perseverance usually paid off.

"Here we are," Lynne said, breaking into my reverie of disasters.

I looked over and nodded. I hadn't even noticed that we'd parked and walked up to the gate.

"You've always pulled through before, Volker," she said. "You'll do it again this time."

So, I went to Memphis and took my boards. It was a three-hour trial during which I was examined by five neurosurgeons and one neurologist in three pairs, one pair an hour, and each pair concentrating on a different area of knowledge: the spine, the brain, and neurology. In each case, the examiners presented a scenario, asked me to make a diagnosis, explain how I would manage it, and how I would deal with complications.

As it happened, Richard Fraser, an attending I had worked with while a resident under Stein, was one of the examiners that day. Because of our connection, he wasn't one of the neurosurgeons examining me, but I saw him at the hospital in Memphis.

Once I got home, I was filled with trepidation. I knew it took a couple of weeks to get the official notification, and I knew that I wasn't supposed to sneak around and try to get those results ahead of time. But, I couldn't stand it, so after three days I called Dr. Fraser and left him a message. I didn't want to put him into the position of doing something wrong, but he knew what I wanted to know. He called me back. "Tell your mother that she didn't raise a dumb kid," was all he said. It was enough.

CHAPTER 8

BRAINSTORMS

*A failure is not always a mistake, it may simply be the best one can do
under the circumstances. The real mistake is to stop trying.*
B.F. SKINNER

I WAS A BONA FIDE, board-approved brain surgeon. Ten years had
passed since I'd graduated from med school in 1971, five or six since, as a
resident, I'd hunched in the cadaver lab for hours doing brain and spinal
cord dissections. I remember how cut and dried—how clinical —that
task became, how the novelty of handling a brain, whether a postmor-
tem one that is pink and fresh, with the consistency of a ripe peach, or a
grayish, firmer one that had been fixated in formaldehyde for a couple
of weeks to the consistency of rubber, well, how that novelty did wear
off over time.

But I also remember how it was to hold a human brain in my hand
the first few times: to place it on the surface in front of me, detach the
brain stem and the cerebellum, as if I were breaking off sections of a
cauliflower, and then make half centimeter slices that I laid out on a tray
the way you would pieces of pound cake or foie gras. A great sense of
responsibility and respect welled up in me because I knew what a phenomenal organ that mottled grey piece of matter was, and I knew it had
contained the essence of a human being, one who had had the grace to
allow his brain to be used to help others.

Now all the brains were still in the heads of living, breathing people.
Of vulnerable patients for whom I was the sole hope of continued existence. How to contain the gravity of such responsibility knowing that I
could never save them all?

Debbie was in her early thirties. Suffering from a crippling headache,
she'd been referred to me by a family doctor in the mountain town of
Payson. He'd sent similar cases my way, and by the time I met with the
patient and her husband for a consultation, the arteriogram I'd ordered

confirmed the worst but also predictable cause for the headache: a giant aneurysm arising from the blood vessel called the anterior communicating artery.

A brain aneurysm is a bulging, weak area in the wall of an artery that supplies blood to the brain. In many cases, it causes no symptoms and goes unnoticed. But often the aneurysm ruptures, resulting in what is known as a subarachnoid hemorrhage. Imagine a patch on a tire. The patch is weaker than the surrounding rubber, and when pressure builds, it bubbles out. If the pressure is not relieved, the patch gets thinner, reaches its limit, and pops. Before that happens, some form of neurological deficit triggers a red alert: double vision, blindness, or—most commonly—the most severe headache the patient has ever had. It also sets a short timer on the clock. The only course of action is to get the patient into surgery as soon as possible, before blood rushes into the brain like the air from that blown out tire. When that happens, the best outcome is a stroke. Often, the outcome is death.

How does one relay this news to a person who, days before, was busy with life? The case was obvious to me, but not for my patient, who had barely had time to absorb the idea that her life was suddenly on the line. She sat grim and alert in my office, squeezing her husband's hand and hanging on my words. Those words were necessarily stark. "This is a very serious condition," I told her. She nodded. "If you don't do anything, the aneurysm may rupture and you have a good chance of dying. But I can get you admitted to the hospital immediately and into surgery tomorrow."

The case was risky, not only because of the size of the aneurysm—a giant aneurysm is defined as one greater than 2.5 centimeters in diameter—but also its location. If you took a sharp stick and pushed it through the bone at the top of the woman's nose, to a depth of three or four inches, the point of that stick would reach the spot where this woman's blood vessel had ballooned up into a ball of blood the size of a small plum. That meant that to get to it, I'd have to navigate through the sylvian fissure, a major pathway beginning behind the eye and running between the frontal lobe and the temporal lobe. Like all the fissures in the brain the natural pathways that appear in a picture as lines dividing the brain into the two hemispheres and subdividing down into the different lobes and sections the sylvian fissure is not clean-cut, not like a canyon running through a mountain. It's a fold in the grey matter that is sticky with arachnoids, spidery web like membranes that I'd have to separate as I went in.

The usual team had assembled in the OR: the scrub and circulating nurs-es, the anesthesiologist, and a sixth-or seventh-year neurosurgery resi-dent to assist. The patient had been prepped. She lay on her back, eyes taped shut, endotracheal tube snaking from under the patches of surgi-cal tape over and around her mouth. Metal prongs, sticking out from the arms of a semicircular brace under her head and screwed into her tem-ples, held her head rigidly still. A three-inch swath of shaved scalp ran just behind the hairline on the right side of her head, from the midline at the top of the skull to above her ear. (Since most patients are right hand-ed / left-side dominant, the incision is usually made on the right side, the non-dominant side.) The line of incision had been marked in blue.

The resident made the first cut. Using a number 15 blade he made a smooth incision down to the cheekbone, or zygomatic arch, just above the opening of the ear. He peeled back the scalp and its pad of fat with a cold scalpel, and used a tool called a periosteal elevator to gently separate it from the bone. "Good and clean," I said, fastening the flap of scalp to the cheek with a series of clamps called "fishhooks." "Now for the burr holes." Using a hand burr called a Hudson drill, he bored three holes into the skull, evenly spaced along a circular trajec-tory above and slightly to the left of the patient's left eye. He then took a stretch of wire called a Gigli saw, threaded it through one hole and out another, and, holding the two ends, sawed out a small saucer of bone. Only one last barrier to the brain remained, the dull gray sheath of dura. In moments, this had been peeled back too.

Performing a brain operation as a young resident under Ben Stein at Tufts New England Medical Center, mid 1970s. Notice the absence of a microscope.

With the brain exposed, it was time to forge a path into the interior. I swiveled the microscope into place. Swathed in clear plastic, lenses of different sizes protruding at various angles, it hung—powerful, precise, ungainly—over the pa-tient's head and upper body like lab equipment still in its packaging. Positioning my eyes in front of the eyepieces, I tested the mouthpiece, clamping it between my teeth and adjusting the view—the mouthpiece would allow me to command the microscope without moving my hands from the surgical field. I took up a pair of microscissors and carefully began to dissect the sylvian fissure.

When a surgeon gets to this point, he understands at a gut level the reference to spiders in the term "arachnoid" matter. Through the microscope the grey web of membrane lay semitransparent and sticky over the fissure. On either side of it the temporal and frontal lobes quivered like Jell-O. Beneath it, the cerebral arteries pulsed in a dense architecture of walls and compartments and cisterns. For an hour and a half I split and separated, split and separated, tunneling deeper into the fissure, advancing slowly along the path towards the aneurysm. At last I reached it.

The aneurysm was embedded in dense tissue, fat arteries pulsing around it. It was so swollen, the artery wall that had bubbled out to form it quivered at the limit of containment. A little nick and it would burst. I continued dissecting around it until it was free from the surrounding tissue. At length, the neck connecting it to the main artery was visible and accessible. "There it is," I said. "We'll need a curved Yasargil on this one, 15 millimeters." I gingerly positioned the tapered blades of the large spring-coil clip over the aneurysm neck, adjusting a hair here, a hair there, finally releasing the clip applier. The clip snapped, sealing the aneurysm off from the artery. With no more blood feeding it, the aneurysm would thrombose—solidify and shrink up. The size of it still concerned me, though. I pricked a needle into the fragile ball and aspirated some of the blood into a syringe. Deflated, it held no more threat.

We closed up, reversing the order we'd followed. Once the retractors were removed, the sylvian fissure closed back in on itself. We closed the dura and replaced the bone flap, wiring it to the surrounding skull. Finally we unpinned the scalp and pulled it back in place, layer by layer, starting with the galia and coming at last to the skin. Sutures along the line of incision completed the procedure. In months, after the hair had grown back, the scar would be invisible.

A successful operation pumps a jolt of euphoria though a surgeon, no matter how controlled we are. I was satisfied that the aneurysm had been physically separated from the parent blood vessel. But the final test comes in the recovery room. When the patient woke up, I went to her. The purple, sutured incision followed the shaved curve of her temporal lobe in a wide "C" down to her earlobe. Her left eye was swollen and bruised. But her vital signs were steady and when she saw me, her lips curved in a weak smile.

"Am I still here?" she said.

"It certainly looks like it," I said. "I'm going to ask you a few questions to make sure. Who are you?"

"Debbie," she replied.

"Who am I?"

"Dr. Sonntag."

"What hospital are you in?"

"Saint Joseph's."

"Good job," I said, and left to share the good news with her family.

Debbie's was the most memorable good outcome of my early days in Phoenix, when I was still primarily doing brain surgeries. Though I had more good outcomes than bad, many of those are a blur now. What a surgeon doesn't forget are the disasters.

I'll call one of these Michael, a pleasant, young black man in his mid twenties who, one day out of the blue, woke up with decreased hearing in one ear. It worsened. In a matter of days he could hear nothing out of the ear. He saw a family physician, who promptly ordered a CT scan and referred him to me. The next thing he knew he was sitting in my office with the films in hand.

Hearing loss typically indicates damage to one of the cranial nerves, the CN-VIII, and such damage usually means a tumor. I clipped the image to the viewer and there it was, in the back of the brain, in a space between the cerebellum and the pons known as the cerebellopontine angle, or CPA. Perhaps 2 centimeters in diameter, against the dark symmetrical Rorschach image of the brain, the tumor stood out round and white as a Ping-Pong ball.

I can't remember the conversation with the patient—except for some banter about where we had each gone to high school. But as with Debbie, I am sure I explained to him the seriousness and urgency of the situation. And though I didn't mention it to him, I couldn't help remembering a similar case I had assisted Ben Stein on near the end of my residency. That patient—a middle aged man—had presented with hearing loss and dizziness. It had been a grueling case. Stein got the tumor out, but the patient was in ICU for three weeks and then on the floor for weeks after that. He developed all kinds of complications, including pneumonia, and never did make a good recovery. Still, such an outcome could well be a good one for this young man, compared to the other way it could go.

A compactly built chief resident named David Pootracal assisted me. To access the cerebellum, at the back of the skull, we positioned the unconscious patient in a sitting position, using Mayfield tongs at his forehead for stability. I went over the game plan. At this point it would have been standard practice for me to leave—ordinarily the resident opens up the skull, only notifying the attending when the brain is exposed. But I had a bad feeling about the tumor and stayed.

David opened up the skull—making the incision halfway between the bump at the bottom of the skull that is known as the inion, and the mastoid, which is at the base of the skull behind the ear—and pinned the skin back with retractors. As with Debbie, the aneurysm patient, he bored two or three holes, inserted the Gigli wire and sawed out a plate of skull, then nipped away additional pieces with a long-bladed, bone-cutting clipper called a Kerrison punch. This time, though, the nurse tossed the pieces of skull in a basin of water to be discarded later; there was so much muscle back there that if Michael survived, he would be left with just a shallow indent.

As soon as the lower back section of skull was off, I cut through the dura, peeled it open and pinned it back. The cerebellum was now exposed. Deriving from the Latin for "little brain," it is far smaller than the cerebrum, comprising only about one eighth of the brain—the size of a small orange. But visually it's not terribly different from the cerebrum: it has two hemispheres and a highly folded surface, or cortex, and it's just about as pink and Jell-O-like as the cerebrum. What is different is that in the cerebellum, you have to deal with the eight cranial nerves that radiate out from the brain stem—which is anterior to the cerebellum—and anchor themselves to holes in the skull. Taking the tumor out without damaging the cranial nerves is painstaking work.

The first half hour went well. My eyes glued to the eyepiece, I carefully retracted the cerebellum and slowly dissected my way into the narrow space of the cerebellopontine angle. Separating, pulling, prying, I reached the tumor where it sat white and pasty in a small pulsing pocket. Now for its removal. A tumor is not a solid mass that a surgeon can simply dislodge and pull out like a marble. It's more the consistency of feta cheese, breaking apart under the pressure of extraction with forceps and suction tube. Still, all that remained was to maneuver chunks of it into a position where we could trap it at the end of the suction and pull it out.

This is where the situation went south. I managed to prod and suck a piece of the tumor out of its nest. I only needed time to get it all. But the brain suddenly swelled like a balloon on a pump. The tumor that, a moment before, had been in reach was eclipsed by a firm, fat, reddish sausage oozing along the line of incision. "Malignant brain edema," I said to Pootracal. "Get those retractors out of there." I pulled back from the microscope for a brief Sisyphean moment. But there was nothing to do but keep picking at the now dangerously swollen cerebellum. I bent to the task again.

For two or three hours I struggled. I dissected pieces of the cerebellum itself, sucking it out piecemeal to clear the way back to the tumor. But just as the prize was in reach—just as I managed to suck out another piece of the white mass—the engorged brain pushed its way in again. I couldn't retract the brain again—I was afraid I might have retracted it too much, decreasing the venous drainage and leading to the swelling in the first place. And I knew that just in front of it, the delicate brain stem lay, controlling Michael's major life functions. If I damaged that, it could instantly pull the plug on his breathing. On I worked, sucking out pieces of the young man's brain. I finally called it quits about three o'clock in the afternoon, after a final fierce struggle to close the skin and suture him up.

The next days were tough. Hooked up to a respirator, Michael's chest rose and fell; his heart beat; he was warm to the touch. At first the Electrical Encephalogram (EEG) showed some brain activity. Still, Michael did not wake up. On the fourth day, he was removed from the respirator during the apnea test. He stopped breathing. He was declared legally, clinically, and measurably brain dead after meeting the rigorous standards set out by the "Determination of Death Act," just passed in 1980, for "irreversible cessation of all functions of the entire brain, including the brain stem."

When I went back to my office later that day, I couldn't stop thinking of the pleasant young man who had come to me in the hope that I could save his life. My failure to do so overwhelmed me. I was painfully conscious of the fact that somehow I had been the conduit to his bad outcome. I couldn't shake the feeling that I had missed something. Had I retracted the brain too much? Is that what had caused it to swell? I stood up, took my wallet from my pocket, and laid it on my desk. Then I opened a drawer and took out a slip of paper. I wrote Michael's name on it, folded it, and tucked it in my wallet. There it would stay for the next thirty-five years.

It wasn't a tumor that struck Melissa, a young woman in her twenties, but an equally nasty growth called an arteriovenous malformation. I'd had a good number of such cases during my sixth postgraduate year (PGY-6) as chief resident. Stein was always the attending on those cases, and all I had to do to know we'd be in marathon mode was to look down at his feet. If he wasn't wearing socks, I knew it was going to be a long haul.

Then as now, I would have seen the angiogram, seen how the usual lacy network of arteries branching off and subdividing into the sponge-like capillary bed—that in turn gives way to individual capillaries that join and form veins to carry blood away—how this elegant web had been invaded by a bloated, wormlike tangle of blood vessels. As I gained experience, my job was to go in under a microscope and control each blood vessel, one at a time, coagulating it with electronically heated forceps and snipping it with the microscissors. It was a mess in there, and the knowledge that this kind of abnormality could often be fatal added to the pressure.

The nest of vessels, called a *nidus*, is extremely fragile and prone to bleeding because of the abnormal direct connection between the high-pressure artery and the low-pressure vein. Imagine trying to untangle a knot of long pasta tubes that was all that was keeping water from a fire hose from passing into a garden hose. Now imagine doing this with tubes whose diameter is measured in micrometers. Even without a stethoscope, I could almost hear the *bruit*, the rhythmic, whooshing sound caused by the blood raging through the arteries and veins.

The resident had opened up the skull on the upper left side of Melissa's head and pulled back the dura before I arrived. Now, eyes to the microscope, he dissected down until the mass was visible. A fat feeding artery pumped into it from above. A draining vein emerged from below with its sluggish flow. Engorged blood vessels threaded through the glistening scarlet knot like malevolent black worms.

"Go to it," I said to the resident.

One by one, he isolated the surface vessels, positioned the long, bayoneted bipolar forceps on either side of them, and with a light step on a foot pedal snapped shut the heated forceps, coagulating the vessels with a small sizzle before snipping them. For the first hour he labored at it, while I manned the suction tube. Then blood began to seep into the field.

This concerned me but it wasn't unusual. Under Stein—he'd been a real stickler about blood control—I'd been trained to deal with it.

"How's her blood pressure?" I said. "High or low?"

The anesthesiologist answered. "It's normal."

"We're getting too much blood here. Bring it down."

I took over the forceps. It was going to be a marathon. Round and round. Deeper and deeper into the tangled morass. Isolate, coagulate, cut. Isolate, coagulate, cut. The nurse mopped my brow. She slipped a straw under my mask. I didn't so much drink the cool juice as absorb it. Four hours. Five hours. The trickle of blood quickened.

"She's bleeding too much," I said. "Check her coag results."

But the results of her pre-surgery screening didn't tell me anything I didn't already know. I could see what was happening; her blood wasn't clotting fast enough.

"Let's get some sponges in there," I said. The nurse passed me the contonoids soaked in a clotting agent called thrombin. They immediately flushed red in the wound. "Gel foam," I said, thinking this next thrombin-based solution would work. It too only absorbed the blood without stopping it. Finally I tried applying pressure—tampenade it's called—wetting a sponge and pressing it against the bleeding vessels. That didn't help. For hours I struggled to get to the core of the knot while blood seeped relentlessly into the cavity, like water trickling into the cabin of a sinking ship. When, hours later, the bleeding did slow down, she became unstable. Her pulse accelerated. Her blood pressure dropped.

Like a contagion, a feeling of futility passed among us.

The resident had stayed the course with me, flushing the wound, suctioning the endless flow. I looked at him, nodded for him to make the call.

"We'd better close up," he said.

"I think you're right," I agreed.

We closed the wound. We didn't replace the bone flap.

For the next three or four days, I saw Melissa twice a day. The hours were limited in the ICU, but my visit coincided with the family's visit on a few of these occasions. Mostly I remember a room full of women: her mother, sisters, friends, holding vigil by her side. The scene played itself out. The first day, the hope against hope that she would stabilize and wake up. The second day, still comatose. The third day the EEG still showed some function, keeping the doomed flame alive. The next day, the brain showed signs of swelling in reaction to the blood in it. Blood outside the vessels irritates the brain tremendously. And if the brain swells, since it's in a closed box—the skull—there's only one thing it can do: squeeze—or herniate—against adjacent structures, in this case the brain stem. As with Michael, we ran through the tests, finally doing the apnea test. And like Michael, Melissa never woke up from the surgery.

Later in my office, a painful regret welled up in me. I pulled out my wallet and removed the piece of paper on which I had written Michael's name months earlier. I unfolded it, took a pen, and wrote Melissa's name under his. Even today, that paper lies creased and dog-eared in my wallet.

CHAPTER 9

BACKBONE

The greatest glory in living lies not in never falling,
but in rising every time we fall.
RALPH WALDO EMERSON

MICHAEL'S AND MELISSA'S surgeries were about a year apart. When I think of them I can still feel the combination of guilt and helplessness that came over me after such outcomes. It was bad enough to come out of surgery physically exhausted and mentally and emotionally drained, knowing that I had not achieved what I had hoped to, but then, going in the next day and the next, and seeing my patient lying there, still and unresponsive, it was like acid dripping on an open wound.

But new cases came, always new cases. As for all doctors trying to establish themselves, it seemed the sun set an hour after I had woken up. I had the two offices I'd kept after Bill Reid left, the one in Phoenix and another in Scottsdale, and though I was getting a few private patients, I was still operating all over the Valley: St. Joe's—where Barrow is located—Good Samaritan, Phoenix Memorial, Scottsdale Memorial, John C. Lincoln. I had made some good professional connections, but there was still no real security in sight. No chance of becoming regular staff anywhere. No progress in joining the teaching faculty at one of the hospitals. We'd taken out a loan from Lynne's father to pay off the mortgage in Ohio, but we still weren't settled into a house we wanted to raise our family in. And we'd barely made a dent in the bank loans we'd taken out.

At times I wondered why I had gone into neurosurgery at all. The odds against success were staggering. I don't remember the statistics back in the 1960s, but these days the attrition rate is astonishing. Figures at the time of this writing show these numbers: 43,315 students apply to med school; 17,759 get accepted; 16,100 graduate. Out of those 16,100 graduates, 310 apply for a neurosurgery residency, for which only 171 are accepted. That means that out of the initial 43,315 medical school applicants only about .4 percent ever becomes a neurosurgeon. The rates are similar for a cardiovascular surgeon. Either way, you can count on six or seven years of training after med school.

In my case, people sometimes assume that my father's brain abscess had something to do with my choice of career. Certainly seeing the disease rob him of his vitality, his profession, and even his country affected me deeply. And it's true that I have carried the image of that jagged scar on his temple in my mind since childhood. But I think other factors weigh much heavier in the choice of what is not simply a career but a vocation. Scratch the surface of a hundred neurosurgeons, and I think you'll see some distinct similarities in character and personality, if not in their formative experiences. (Much later Ben Stein whittled it down to one distinct trait shared by neurosurgeons: "We're all a bunch of wild men and egomaniacs," he said.)

I'm not sure how much of a wild man I am, but when I look at the kind of kid I was, even before I came to America at the age of twelve, I see certain tendencies that supported my later endeavors. One was a positive responsiveness to discipline.

Discipline was still very much enforced at school and at home in the Germany of the early 1950s, despite any fallout from the Nazis' manipulation of the German value of obedience to authority. At school, for example, there was Mr. Sprenger, my teacher from grade one through grade four. In his thirties or forties, he always dressed in the requisite severe woolen suit and tie. While we students labored over our assignments, he sauntered up and down the aisles tapping a small whip into the palm of one hand—the punishment for cheating. The discipline worked with me. I was not a rebellious boy. Adherence to rightful authority lodged firmly in my psyche and influenced my work ethic, sense of duty, and later political persuasions.

Having said this, when I was young, obedience was to an individual, one's parent or teacher or another adult, not to a state embodied in one leader. My classmates and I were not being trained to be little nationalists. Indeed, in the postwar years, there was a distinct aversion to manifestations of patriotism, reflected in what was absent from the school environment. The front of each large, airy classroom was dominated by a wall of blackboard, and there were several long metal cylinders containing rolled up maps in a corner of the room, but there was no flag to be seen anywhere, and certainly no visual reference to Konrad Adenauer or to the German state in any form. Nor was there any indoctrination through a pledge of loyalty to the state. At school, obedience belonged to Herr Sprenger and his little whip.

At home my mother was in charge of discipline, and I remember balking at her tactics—there was the pantry punishment, and the swift whacks with the back of her hand. I did sometimes silently rebel—I remember fuming in the dark amid cans of oatmeal and flour, and glass jars packed tight with pickled cherries or vegetables, thinking: *She'll be sorry when she comes to let me out and finds me dead with my wrists slit.* But I learned to behave under what was a fairly typical child-rearing philosophy of the time. My mother was simply practicing what every self-respecting *Hausfrau* did. It was a "spare the rod, spoil the child" take on things.

I suppose my mother—Gisela Albrecht Sonntag—absorbed some of the lessons of raising children from her mother, but her upbringing was so drastically different from my brothers' and mine that it's hard to say how much the intervening events had influenced her. Born on the cusp of World War I, the eldest daughter of the mayor of Güstrow in northeastern Germany, she'd had a very happy childhood. She and her younger sisters, Gurtrude, Eva, and Runhilde, and her brother Karl, shared with their parents a grand, two-story house that was so large the East German State later divided it into five or six separate apartments. I never saw that house as a child, never saw it while the Berlin Wall stood, but I knew about the large walnut tree on the grounds; at Christmas, my grandparents always sent us a bag of walnuts from that tree, the shells into which my father bored small holes and strung up on our Christmas tree.

My mother was strikingly beautiful in her youth, and had a refined if dramatic sensibility. In a portrait taken when she was eighteen years old, wide-spaced eyes gaze out frankly from beneath a short, gleaming boyish haircut. She has a slightly rounded, serious face, distinguished by a wide mouth and a long, straight nose. Her neck, encircled with a single stand of pearls, rises like the stem of a flower from an upturned collar. She seems to have fully embraced the modern *zeitgeist* of her times, taking herself off to Berlin to teach gymnastics when she was in her early twenties.

Those were the heady years of the Weimar Republic, a time of economic uncertainty and increasing political extremism but also one of renaissance in the arts, when Germany became famous for its cafe culture, its Jazz clubs and rising film industry. From 1919 until 1933, when my mother was on the cusp of her womanhood, the modern design and architectural movement called the Bauhaus swept across Germany and influenced capitals across Europe and the Atlantic. In 1927, when my mother was thirteen years old, Fritz Lang produced one of the most famous films of all time, *Metropolis*.

In hindsight, we shudder to think of the chaos that was about to engulf Europe and the world. But what did my parents know? How political were they?

In 1937, at twenty-three, my mother married my father, who was ten years her senior. My older brother Gunther arrived in 1938. I don't know when it happened, but my father moved the family further east to Graudenz, where he established a dental practice. On September 3, 1939, Hitler's Panzers rolled down the cobblestoned streets of that very town; the cheers of jubilation from the minority German population are well documented in a famous photograph (though the horror of the Poles is not). Their lives seem to have retained some normalcy; there is a photograph of Gunther's baptism, pictures of a beach holiday around 1940—Gunther is just walking—and snaps of my parents in a train station. And there's one of a military band playing in the main square of Graudenz after the Nazis came to power.

It is a well-composed photograph with a stamp on the back reading FOTO-HAUS-WALESA, GRAUDENZ, HERRENSTR, 34. In the foreground to the right can be seen the backs of young men in uniform standing at attention while in the middle ground the band members, also in uniform and following the sheet music open before them on stands, play for the assembled crowd. To the left can be seen the townspeople who have come out to watch, mostly unsmiling men and young boys. Behind them, in sharp relief, stand the facades of buildings on a city street adorned with banners emblazoned with the Nazi swastika. It is a representative historical photograph of the time, but what is most telling about it is one, small detail: Blurred but visible on a first-floor balcony above the crowd is a thin man leaning with his hands on the balustrade. He is the only person viewing the scene from above. Only if you take a magnifying glass or press your nose right up against the photograph can you see the tiny Nazi flag suspended from the man's balcony. That man was my father, Heinz Sonntag.

It was my cousin Heinz Peter who told me the story that goes with that photograph. The townspeople living or doing business in the buildings fronting the street had been instructed to display the Nazi banners from balconies and flagpoles. The section of street visible in the photograph shows four such banners. My father, occupying the apartment above the square, did not wish to defy the instructions outright, but neither did he wish to show his support. He got the smallest flag he could find and hung it up. The gesture was seen for what it was, and my father was reportedly carted off to jail for the night.

I didn't know any of these stories until years later, after both my parents had died. It was only then that I opened the photo albums my parents had kept and went through their pictures, (among them a portrait of my uncle Karl in his *Luftwaffe* uniform; he was shot down sometime before my birth in 1944). My father never talked about the war or his life before it, and my mother only brought it up one beautiful spring night in 1986, when she abruptly sat up from her deathbed saying, "There's something I want to tell you."

What I did know as a child was that my mother was strict, reserved, and proud, and that my father was gregarious and enthusiastic, a can-do kind of man who loved to laugh and socialize. Both influenced me, but my mother was the disciplinarian. Between her and my teacher, life was an orderly, predictable affair. It was not that my father did not wield authority in the house, but he was at the office near the *Markeplatz* most of the day, and my mother was inherently more of a taskmaster than he was. School only ran until one o'clock in the afternoon. Then I went home to the midday meal of *Schmaltz* on bread and, under my mother's watchful eye, did my homework. There was never any question of the order of things. Work always came before play.

That kind of early external control kept me in line. But there is also personality. Both my brothers shared my early environment, but we all took different paths. My older brother Gunther was brilliant, and though he went through a rebellious spell around the time we immigrated, he became a successful banker. My younger brother Rüdiger didn't like the rigors of school but he was gifted in woodworking and had a good career in the army. There's a line from the nineteenth-century author and educator Henry Van Dyke that hits the nail on the head, I think: "What we *do* belongs to who we are; and who we are is what becomes of us." What I did was to explore as much as I could within my family's limited means. Once we'd arrived in America, with public school and the uniquely American customs of organized sports and after school programs at community centers, that exploration turned out to be a lot. My assimilation to American culture was under way.

That assimilation first took the form of being thrown into a sixth-grade classroom at the end of the school year and treading water until my language skills picked up. Thanks to Annette Funicello and the Mouseketeers, and the local Wallace and Ladmo children's television program, by seventh grade I was, if not fluent, functioning in English.

Then there was my awakening to recent history and how it played out in my own life. We were immigrants in America, and like all immigrants, we thrilled to Hollywood's depiction of the cities and the Wild West and the cars

Entering the Soapbox Derby ca. 1958. Through school and community activities, my assimilation into American life was swift.

and all those other wonders. What we never discussed was the discrimination immigrants faced, although my parents must have known that as Germans we were potential targets. One day, my mother was forced to address the dark side of my German identity.

It was that first year in Phoenix. My parents had made friends with another German couple who lived in at tract home next to the trailer court. I had to take something to them and was chatting with them in German as they walked me out. We were in the carport when a boy my age passed by. I had seen him around but didn't know him. The boy looked at me appraisingly before addressing me:

"So you're German," he said. "You're one of those Nazis."

Nazis? I didn't know what to make of that. I went home and found my mother in the kitchen cooking dinner.

"Mutti, what is a Nazi?" I asked her.

At first, I thought she hadn't heard me. She busied herself for a moment and then turned to me.

"Well, Volker," she said, "the Nazis were the people in charge of Germany when Germany was fighting the rest of the world."

That explanation satisfied me for a while. I knew nothing about World War II. That period of recent history had not been covered when I was in school in Germany. I knew from our experience on the ship that something bad had happened, and that was why we were with all those other displaced people. But I think I attributed our particular situation more to my father's illness than to some larger forces.

At Central High School I became more Americanized, but I knew some students still saw me as a "German kid." There had been a teammate's response to a casual remark I'd made my sophomore year, while practicing cross-country. "Okay, you'd better watch out, I'm going to beat the pants off you," I said at the starting line. "You guys weren't able to beat us in two world wars," he threw back at me. "What makes you think you can do it now?" Another teammate addressed me as the "Aryan on the team" in a yearbook message.

Those boys probably didn't give a second thought to what they had said, and I took it in stride. But by high school, I knew what it meant to be German. I can't peg a particular memory to the first time I heard about the Holocaust. I remember only a horrified wave of disbelief that anyone could conceive of such atrocities as gas chambers and ovens and concentration camps. And while I experienced no overt rejections or challenges to my Germanness, at times I would be left wondering about an encounter. Did that Jewish kid tell me I couldn't enter his house because it was the Sabbath, or because I was German?

I had no answers to such questions. But the gradual realization of my larger history as a German and a burning desire to prove myself drove me to push ever harder. But it was even before high school that I encountered a man who would imprint on me a value that would stand me in good stead all my life: hard work.

One day about a year into our new lives as Americans, I was mowing the front lawn as part of my job for Mr. Beuchel when a beautiful, robins-egg-blue Ford station wagon swung around the curved drive in front of our house at Mr. Beuchel's trailer park. It took me a moment to realize the woman behind the wheel wanted to talk to me.

"Hey kid," she said, waving me over from the frame of her open window, "you want a job?"

I looked in the back of the wagon. It was loaded with small paste-board boxes the color of oatmeal, each of which had a stamped pattern of chickens running across the front and the words: FRESH EGGS.

I think I shrugged a "why not?" However it happened, that was how I ended up working on Mr. Reynolds's chicken farm.

Mr. Reynolds was a small-muscled, wiry man in his late forties or early fifties, a no-nonsense kind of person who didn't waste words. It was his wife who had poked her head out of the window to offer me a job. When I trailed around him that first day, all I was thinking was how generous he was to have offered me thirty-five cents an hour, a whopping ten-cent raise from what Mr. Beuchel had paid. I had no inkling that the job I had just signed up for would be one of the most formative experiences of my life, or that Mr. Reynolds would become my model of how a man should conduct himself.

Seven long rows delineated Mr. Reynolds's chicken property, each with four giant coops. Each coop housed about four hundred chickens. It would be my job to swab out the long troughs that ran along the ends of these coops, twenty-eight of them in all.

I followed Mr. Reynolds down to the end of the first coop, nervously meeting the beady gaze of the chickens lined up on planks stained white with droppings. The birds clucked and fussed, flapping their wings and sending small dust devils of feathers and dander into the air. The trough there was filled with brackish poop-and-algae-clogged water. At one end was a spout where the water flowed in, and at the other end was a lever attached to a thin chain that disappeared into the mucky stew. The chain was attached to a rubber ball that acted as a stopper at the bottom of the tank. I would have to pull that ball out, let the water drain, and then scrape the muck into a gelatinous mass that could be scooped up and discarded into a dump outside the coop. Then I would scrub down the trough and replace the ball. Twenty-eight times a night.

I couldn't believe my luck to have landed such a good job.

The job was not just something to give me a little pocket money. Foul as the work was, I depended on it for clothes and books and other necessities, since my parents could barely afford food and rent, let alone repayment on the loan for the trip over from Germany. A document dated July 9, 1957 shows that they owed $1167.34 upon our arrival in the U.S. The agreement with the Lutheran Refugee Service had stipulated that my parents pay fifty dollars a month, and that the entire debt was to be paid off within two years. They had managed to keep up with it through the first summer, fall, and winter, but in February, my mother was forced to write to the Service and ask for a grace period. My father was unemployed again, and there was no way to come up with the money, even though my mother had secured a job as a secretary for a dentist named Dr. Bobo. When they did resume payments a month or two later, all they could afford was twenty-five dollars. That soon went down to twenty dollars, and before the year was out, fifteen dollars.

Things got worse. Around the time I started at the chicken farm, my mother had a falling out with our sponsor's wife, forcing us to move from the trailer court to a place we called the "Shed." No bigger than seven hundred or eight hundred square feet, it had two cramped bedrooms and a kitchen so small we had to push the table up against the wall and sit squashed together on the three remaining sides. The only good thing about that house, apart from the cheap rent, was that it was closer to the chicken farm.

The first house had had a working swamp cooler; the one at the shed labored to exhaled a lukewarm mist and made such a racket that you could hardly sleep at night. The first house had sheltered under shade trees; this one was exposed on three sides to the unrelenting sun. The first house had had grass and orchards all around; this one had a parched patch in the front that looked like a dried up, yellow carpet. It was as if that house were mimicking my father's stymied situation. Neither was flourishing.

My mother was hardly flourishing either. She was dropping weight at an alarming rate. That sent her anxiety in the opposite direction. She became so jittery at night that she couldn't sleep with my father anymore and had to take my bed, sending me off to sleep with him.

I didn't realize the stress she was under. I didn't notice how gaunt she was getting. It was only much later, looking at photographs from that fall and the following year that the seriousness of her condition became apparent to me. In one picture, her figure is so flat it looks as if you could fold her in half like a piece of paper. In others, her sleeveless summer dresses reveal long, spindly arms with little definition of muscle from her shoulders to her wrists. And her face—it had always had a slightly asymmetrical quality due to a weakness in the muscles around her left eye. Now whatever was making her drop weight had narrowed that eye even more, giving her a skeptical countenance, as if that eye reflected her disillusion with the way our lives were going.

They were not going well. My father's inability to get and keep a job was taking its toll. My mother was becoming the authority figure in the house, and my father resented his loss of control over what was happening to our family. At night, their voices mounted behind the wall. It was the same argument over and over:

"Heinz, we have no money for that!" I heard my mother say more than once when my father brought home a bottle of the local A1 Beer.

"*Mein Gott* Gila," he said, "it's forty-nine cents. You're going to deny me a forty-nine-cent bottle of beer?"

"Forty-nine cents is forty-nine cents. You have to earn forty-nine cents to spend forty-nine cents."

My father had no comeback to such comments. How could he justify in his mind that he was depending on his wife, that she was supporting the family? He grew increasingly bitter.

I kept myself busy. At the chicken farm, Mr. Reynolds was giving me more responsibility, teaching me everything he knew by example. I learned how to "candle" the eggs in the mini-barn next to the house by watching Mr. Reynolds hold the egg up to the light to identify if it had been fertilized or not; how to discard the fertilized ones and put the others on a conveyor belt that separated them by size; how, when the chickens had reached six months of age, to round them up in a corner of the coop, fence them in, and one by one, open the beaks with my finger and place the upper one under a hot blade that cut it off and cauterized it at the same time; how to grapple a chicken down, lift it up by the legs and hand it over to Mr. Reynolds for its annual vaccination; how to do the filthiest work imaginable without complaining.

It was by example that this taciturn, dust-caked, man's-gotta-do-what-a-man's-gotta-do chicken farmer taught me how to be a man. When the monsoon rains came late in the summer, I never heard him complain about the reeking smell that emanated from those coops and hung in the damp air like a putrid cloud waiting to assault our noses. When the time came in the winter to wrap large swaths of dirt-and-poop encrusted burlap around the coops to keep the chickens warm, he went at it like a housewife hanging out a clean sheet in a summer breeze. By the time my first year with him rolled around to summer again, there was no question that I would stay on when I started high school in the fall.

I can't remember the moment I decided to become a doctor, but I do remember writing to my boyhood friend Frieder back in Germany, that first year in America, that I intended to do so. By high school, I knew medicine was what I wanted to do with my life. I took the usual math, science and English classes in high school and for my foreign language elective, I opted for Latin and found that it resonated with my personality, just as math did. Its rigid structure made sense to me the same way making my bed every morning and keeping my room neat made sense. I enjoyed the clean rules of the grammar and the way adjectives changed depending on how you used them. Latin seemed a much more defined language than either German or English, and though I don't think it helped me later on, I never regretted staying with it all four years.

History and literature got my guns going too. One teacher who influenced me was my English teacher, Mrs. Boyle, a warm, older woman with short, dark, wavy hair and a thin, pleasant smile. I sensed that she took a personal interest in me—maybe a good teacher makes all students feel that way—and I respected her greatly. We read *Catcher in the Rye*, *To Kill a Mockingbird*, and *Lord of the Flies*, books that challenged me to think about what it meant to live an authentic life. The one big thing I took away from Mrs. Boyle was the idea that it was important to do good in life. That is a simple message, but it's the critical one.

I also knew, inherently, that physical fitness was essential. I had played soccer from my earliest boyhood in Germany, but sports in the US were much more organized and competitive. I went out for the usual athletics: football (my mother put the kibosh on that after my freshman year), basketball, and track. I didn't know what I was good at in track, but I ended up running middle or long distance and eventually ran the mile. Now looking back, it seems I was running figuratively too, towards my identity as an American, towards my future in medicine.

Now, here I was, nearly ten years after giving the convocation address for the first graduating class at U of A medical school, still struggling in private practice, still pursuing a foothold in the relatively closed neurosurgical circles in Phoenix. I tried to keep a healthy perspective on my career, (though my mother-in-law's words from a few years earlier frequently echoed in my head: "Are you ever going to make any money at this?" she had asked during a visit with Lynne and me in Boston). I reminded myself that in light of my parents' struggles and my own early experiences, things were not going all that bad. Take away the mounting debt, the redlined balance sheet for my practice, and my failure to make some crucial connection, and we were doing just fine.

What I didn't realize was that I had already made a critical connection, and that it was imperceptibly pulling me in a direction that would define my career. As a way to keep connected, as well as to keep my own knowledge up to par, I had been attending teaching rounds at St. Joseph's Hospital, since my arrival back in town, in pathology, morbidity and mortality, and other topics that were beginning to be required by residency review committees. I had taken this tact (as I had the practice of attending doctors' meetings at the hospitals) hoping that it would expose me to the big players in town—especially the influential neurosurgeons. What I had not anticipated was the importance of connecting with the residents.

During that first year, I had built up a small reputation with the residents at Barrow as a neurosurgeon who could handle certain types of cases. Finding that I enjoyed working with the residents, a fateful if simple idea came to me. Barrow didn't have a strong residency program, my own oral boards were coming up at the time, and I was eager to review critical topics myself. I tracked down the chief resident and suggested that I do Wednesday morning board review rounds for the residents. In that short, decisive moment, I was pulled back into the academic track that I had rejected with Ben Stein, pulled back into a secondary vocation I was destined to fulfill—teaching.

CHAPTER 10

THE WEAK SISTERS
OF NEUROSURGERY

Intelligence without ambition is a bird without wings.
SALVADOR DALI

LET'S SAY YOU COULD remove the skull of a person sitting in front
of you, eliminate all flesh, muscle, and soft tissue, including the spinal
cord, and peer down into the empty spinal column. You would see a
small knob protruding from the front of the top vertebra. If you looked
closely, you would see that the knob was more like a peg, and that it actually originated a bit farther down, at the second cervical vertebra, C2.
This bone is the odontoid process, also known as the dens (it looks like
a tooth) or peg. It's the bone that C1 rotates around, allowing far more
motion that any other level in the cervical spine; without it we could not
swivel our heads. It is also the bone that accounts for a large number
of spine injuries, especially in the elderly and almost always from falls.
Along with a few other features of the spine, it was this bone that, in
ways I could not foresee, began gradually to pull me in a new direction.

It was a very subtle shift. I was a brain surgeon. I had trained long
and hard in my area of expertise. And yet, I was still on the margins.
It was nearly impossible to break into the neurosurgical establishment.
There were three or four local group practices in neurosurgery, but few
opportunities to join one. A handful of individual neurosurgeons cor-
nered the action at the local hospitals, and if it hadn't been for the County Hospital cases and the odd referral, I might have fulfilled Dr. Green's
dire prediction about starving.

If there weren't enough brain cases to go around, there were plenty
of spine injuries. But no one in the neurosurgical community was clam-
oring to take those. The brain was sexy: clipping aneurysms, taking out
tumors. The spine was assembly line work, and not very lucrative at that.
Moreover, unlike today, where a neurosurgeon or orthopedic surgeon can
repair the dens, there was no surgical treatment for such injuries. So when
one or two spine injuries came into the ER on the weekend, the residents

would assign bed rest, put a collar on the patient or do skeletal traction, and wait until Monday to find a surgeon who was willing to take the patients on. Given the paucity of opportunity open to me, I was just the man they were looking for. Soon word got around and even though I wasn't officially on call at St. Joe's, the residents there started saving those cases that for a neurosurgeon were the "weak sisters" of neurosurgery.

So it was that I arrived in the ICU one Monday morning to find Mrs. Merrick, a frail, heavily wrinkled 77-year old woman with thin, ice-blue hair lying in bed with her eyes closed. Her small chin disappeared into the soft collar they'd put on her, and her hospital gown hung on her meager frame as if on a hanger. A resident accompanied me. I'd already seen the notes on the chart, and as we entered the room he voiced what I already knew. "Admitted Saturday noon. Fracture at the base of the odontoid. Prescribed bed rest. They put a collar on her, but she hasn't had any traction yet. Presume she'll need a halo."

I concurred about the halo. It was the usual procedure.

"Good morning Mrs. Merrick," I said.

She opened her eyes. They were remarkably clear and dark, as if they hadn't aged along with the rest of her.

"I'm Dr. Sonntag, a neurosurgeon, and I'll be taking care of you. As you know, you've fractured your odontoid, a small bone at the base of your skull. That has made your cervical spine very unstable. We're going to have to stabilize it by putting you into an external brace."

Mrs. Merrick nodded into her collar, and raised one bony mottled hand slightly before letting it fall. "I know Doctor," she said in a feeble voice. "It's such a stupid thing. It's not the first time. I should have been more careful."

"It's quite common," I said. "We'll get you stabilized and hopefully you'll be home in a couple of days."

It was a scenario repeated dozens of times at thousands of hospitals. The admittance of an elderly patient from a fall or other accident; initial treatment in the Emergency Room; the move to ICU; and referral to a neurosurgeon or orthopedic surgeon. After that, the patient was channeled to orthotics for the fitting of a halo brace.

At St. Joe's, the orthotist was an efficient man named George Mattingly who presided over the cast room with its collection of vests and halo crowns in a range of sizes. The surgeon in charge, I was always present with the patient in the cast room for the fitting, and so I was with Mrs. Merrick. She lay flat on her back, her eyes taking in what she could in her range of vision—mostly the square white ceiling panels above her.

"We're ready to begin Mrs. Merrick," I said. "This is George, the orthotist, and he's going to be helping me make sure the halo fits well and that you'll be comfortable moving in it."

"OK," Mrs. Merrick said. She closed her eyes and her face took on a pained look of resignation.

I picked up a syringe. "First I'm going to give you an injection so you won't have any pain when the pins go in." I injected 1-2 cc's of zylocaine above each eyebrow and about an inch in from the back of both ears. I waited a moment for the drug to numb the skin down to the periosteum level of the skull. Then I nodded to George. He removed the soft collar, selected a crown, and handed it to me. I positioned this ring, or halo, around Mrs. Merrick's forehead, allowing for a finger's breadth of space between metal and forehead. Then I twisted a first temporary screw through a perforation in the ring until the flattened end pressed into the skin above one eyebrow. I repeated the process above the other eyebrow and twice in the back of the skull. Satisfied with the positioning of the halo, I replaced the temporary screws and, one by one, with a regular L-shaped Allen wrench twisted the pointed screws in until they penetrated the skull. Mrs. Merrick's mouth disappeared into a thin, grim line. Throughout the process she exhaled sharp puffs from her nose. I was sure she was not in much pain, but I've never seen a patient who didn't flinch at screws being driven into their skull.

George and I fitted the lambswool vest around Mrs. Merrick's torso. Four black metal rods dangled awkwardly from it, two in front and two in back. Now all that remained was to connect the ends of the four rods to the halo. When we'd finished, we helped Mrs. Merrick on with her gown and supported her while she got up off the bed. After a few flailing attempts, she got used to the inhuman rigidity of the brace. Mrs. Merrick now faced six to twelve weeks in the halo and after that a soft collar during the therapy that followed to strengthen her neck muscles.

Mrs. Merrick was fortunate to have fallen in the early 1980s when the modern halo brace had been around for about ten years. Between the 1930s and the late 1950s, when the halo was first described, any stabilization or reduction of cervical spinal cord injuries, such as traction, was bed based. Long-term confinement to bed brings a host of serious complications, among them pressure sores and muscle wasting. Then in 1959 two orthopedists, Vernon Nickel and Jacquelin Perry, developed a cumbersome contraption that encased the entire torso in a cast-like jacket that rested on the hips. The jacket was modified into the vest and the brace itself was streamlined in the 1970s. Now Mrs. Merrick enjoyed a significantly improved model.

Bed rest and the halo constituted the main treatment for other spine cases as well, and though there were a couple of other procedures depending on the particular injury, there was nothing cutting-edge or exciting about any of them. If the case involved a herniated disc, for example, you could prescribe decompression: strapping the patient to a special table that targets the problem disc and gently pulls the spine apart to allow the disc to pull back in. If it was more severe, and the rupture in the disc had allowed the soft center to spurt out, affecting the nerve root, you went in under a microscope and did a discectomy. That operation was more rewarding than fitting a halo brace, but it wasn't brain surgery.

Not that a discectomy was simple. Today a minimally invasive discectomy for a herniated disc can be done on an outpatient basis, but then it was a major operation, requiring a hospital stay of three or four days and an incision of at least 1 inch to 2 1/2 inches, sometimes 3. I did so many of these surgeries that they blur together now. I do remember one particular patient, though, a middle-aged, morbidly obese man from Hawaii named Pat Kamaka, who came into the emergency room one afternoon with severe pain in his lower back and legs.

I met Pat in the emergency room. He lay on a special hospital bed for wide patients—the EMT had called ahead to alert the ER staff of the special circumstances—and he was breathing heavily through an oxygen mask a nurse held up to his face. A thigh-sized blood pressure cuff aspirated while it ballooned up on one forearm, and the usual electrode and IV wires coiled at his sides. I wondered how the EMTs had been able to load and unload him.

He'd arrived with his diagnostic studies, but even without them his incontinence of bowel and bladder made it clear he needed an emergency discectomy. "You have a ruptured disc in your lumbar spine," I told him, "and there's a severe pinching of the nerves going down to your legs and to your bowel and bladder. I'd like to get you into surgery as soon as an OR room is available. We need to be quick; this is not something that can wait."

It was hardly surprising news to Pat. Given the stress put on his spine, the rupture had been only a matter of time. He couldn't walk at that point.

The resident who happened to be available that afternoon was a lanky, boyish man named Mark Garnett, one of the nicest people I've met. Before we scrubbed, we went over the plan.

" Should be a standard discectomy," I said. "We'll need to make some modifications for the patient's size and it may be a little more difficult."

"Right, his obesity,' Mark said. "What about the increased pressure on his abdomen?"

"That'll certainly increase the venous pressure in the *cauda equina* (the nerves around a lumbar disc), so we'll have to monitor that carefully. We'll also have to make a larger incision than usual, since we'll have to go through deeper layers to get to the disc."

Mark nodded. "All right Chief. I'll follow your lead."

"I think we're set then, Chiefy," I said. At least *you* won't have to stand on a stool."

Pat's herniated disc was in the lumbar spine—the lower back where the spine curves inward toward the abdomen. The disc—which acts like a kind of shock absorber for the spine—was between L4 and L5. MRI images had given us a clear image of the rupture, directing me where to make the incision. A surgeon always wants to cut as close to the problem area as possible, which isn't as easy as it sounds.

Mark and I took a last glance at the MRIs and moved in with the additional nurses that had gathered around the operating table. The anesthesiologist finished intubating Pat, who lay—a seemingly immovable mound—on his back. Within moments Pat was under. The anesthesiologist supported Pat's head while six of us readied ourselves, three on one side and three on the other. Mark, a nurse, and I gripped the webbed handles of a special bariatric roller sheet that lay under Pat's hips and shoulders. We stepped back, leaned back, and pulled. Pat budged an inch. We pulled again. Another inch. As we continued our short bursts of pulling, the team on the other side aligned his legs with his torso and maneuvered his arms, making sure the catheter, IV, and other tubes remained in place. At last, Pat heaved over onto his stomach near the edge of the table. A final pull or two from the opposite side and he was centered. The anesthesiologist positioned his face in the horseshoe-shaped headrest. We all breathed heavily.

Two nurses now draped Pat's backside, except for an exposed 5-inch section of his lower back, which had been prepped with an antiseptic solution. Standing on a step stool for added reach, I pressed the small curved blade of a # 15 scalpel against the swollen flesh and made a vertical incision of $2^{1}/2$ inches straight down the middle of his spine, exposing a thick layer of yellow fat the consistency of cottage cheese.

Mark set the suction going while I dissected through the fat with an electric Bovie knife. Each time I activated the hot current with the foot pedal, the rank odor of sizzling human fat rose like smoke. Slicing and sucking through the slippery mass, our surgical gloves greasy with it, we managed at last to retract the slimy skin. The muscles in my back and arms already ached, but I'd only reached the thick muscle. I continued, separating the muscle from the bone and cauterizing it with the Bovie knife. Some minutes later, deeper in, I called for a Taylor retractor. Mark inserted the second retractor on the muscle, at last opening a neat, roughly diamond shaped window on the vertebral column.

The ruptured disc was not visible yet. What confronted me were the bumpy spinous processes and the sloping lamina of the vertebra. On the other side of that barrier lay the *cauda equina*, a cable of fibers and nerves buzzing with electrical impulses from which, at every vertebral level, a pair of spinal nerves branches out to the body to relay sensory and motor information from the body to the brain. Under the microscope I drilled a small opening in the lamina and dissected through the so-called yellow ligament. Only then did I encounter the ruptured disc.

Now I had to make sure I was at the right level. In this case I needed to be precisely in the disc space between L4 and L5. "Number 4 Penfield," I said, calling for a thin dissector to insert into the interspace. "Let's run an X-ray on it." The image showed the thin blade of the dissector perfectly positioned, like the proboscis of a hummingbird, between the two discs.

I've heard a herniated disc described as a jelly donut with a hole that allows the filling to ooze out. It's comprised of two parts, the fibrous outer ring called the annulus, and the pulpy core, the nucleus. The disc herniates when there's a defect in the annulus. The defect might be caused by the individual's bending funny, as a result of trauma, or at times due to an unexplained rupture. However it happens, when the annulus tears, the nucleus ruptures and can put pressure on the nerve roots, sending shooting pains into the legs, buttocks, and lower back. It can also cause numbness and weakening in the legs.

After retracting the L5 nerve root, there it was, the pulpy nucleus that had erupted, like a condensed ball of moist crabmeat. With straight-angled pituitary forceps I removed the ruptured disc material and dropped it into a dish. Then I entered the interspace, easily removing the disc fragments. After closing up the dura, one last step remained. I dissected a small slab of that yellow fat out from the subcutaneous area and placed it over the nerve root. The fat would prevent the growth of scar tissue. Then we closed him up.

Pat had a decent recovery—after a couple of days in ICU—but it was not unusual for a slew of complications to occur during such a surgery: a recurrent disc where part of what you haven't taken out ruptures again later; a persistent disc where you didn't take it all or you took too much bone and caused instability, or even an inadvertent durotomy—a nick in the dura that the surgeon didn't intend to make and didn't realize at the time. Such a tear can cause a big leak of cerebral spinal fluid into soft tissue outside the dura, forming a capsule called a pseudomeningoceles. Not that your body can't replace the spinal fluid—your body makes $1/3$ of your total supply of spinal fluid every hour. But the capsule can pump up to the size of an egg or a grapefruit, depending on where it occurs and how long it takes for the problem to be detected. Then it's back to surgery for the patient, who has now likely added headaches, nausea, a recurrence of the low back pain, and other symptoms to the original complaint.

Then there were the complications associated with obesity. The layer of fat is a kind of dead space; there's nothing you can suture and it's prone to infection. Moreover, fat people often have diabetes, which, among other problems, increases the infection rate. High blood pressure, slow healing, and prolonged physical therapy were certainly problems Pat had to contend with. But fortunately he had no complications, and he made an excellent recovery.

I wish I could say that I foresaw the coming advances in biomechanics and instrumentation that would make the spine a hot neurosurgical specialty, that I was inspired to improve the experience for patients such as Mrs. Merrick, Pat, and those dozens of discectomy patients. But the truth is, I felt like every other neurosurgeon: the spine cases were a necessity, but I wanted the brain cases. All the challenge and glory lay with the brain. Doing the spine was like being an astronaut but never getting into space.

Still, I was hungry. I had bills to pay. And though as a surgeon I couldn't make any money having patients lying there day after day while I checked if they could wiggle their toes yet or not, I could charge for putting on the halo and operating on herniated discs. That was the reality of it.

My own background is illustrative of what an afterthought the spine was in neurosurgical circles. When I finished med school in 1971, there had been few neurosurgical contributions to the treatment of spinal disorders. Neurosurgeons of my generation were primarily trained in the treatment of intracranial abnormalities, with only passing reference to the spine. For example, the annual review from the 1965 Congress of Neurosurgeons meeting, *Clinical Neurosurgery*, devoted only two out of twenty-eight chapters to spinal disorders, and that was a good track record compared to the review of 1981, none of whose thirty-seven chapters was dedicated to the spine and most of which covered cerebrovascular abnormalities, followed by head trauma and brain tumors. As at the Congress, so in the professional organizations. The major neurosurgical societies and associations of the time all emphasized the cranial at the expense of what is essentially the outgrowth and extension of the cranial—the spinal cord.

Despite this minimization of the spine, I had experienced very complicated spine cases, in particular tumors inside the spinal cord. Those are called intramedullary tumors. An intramedullary tumor is the size of a cocktail sausage; on an MRI scan the lump it creates in the spinal cord like looks something like a mouse in the belly of a snake. Your spinal cord is only about as big as your finger, and once you get in there, you see a slew of fiber tracts that run along next to the tumor. These fibers are axons (nerves inside the cord) that control the bowels, bladder, sex, and the arms and legs. If the tumor were a self contained little nut, you could just pluck it out. But there's never any "plucking out' in neurosurgery, only—as in this type of case—painstaking dissection under the microscope.

A spinal cord AVM was another type of complicated surgery that, though extremely rare, I had had experience with. There was a pilot for American Airlines who came in with a vascular abnormality of this sort during my residency under Stein. I had to cut open the spinal cord to get at the knot of blood vessels. This was about 1975 or 1976, just before advances in neuroradiology and micro-neurosurgery allowed us to better understand the physiology of these abnormalities. In those days, we thought the whole long vein on top of the spinal cord, and its connection to the artery, was the problem, so we simply dissected the vein from the many segments of the cord, separating it from its arterial supply. That was the usual procedure until the end of the 1970s. Now we just clip that abnormal connection between the vein and the artery, deflating the vein. The pilot had a good outcome, though, and it made for a good show and tell in the grand rounds right after that, when I passed that vein around like a kid with a snake at Show and Tell.

If I had to do the spine, those were the cases I longed for: the complicated surgeries that would test me and advance my career. Instead—though some challenging cases came my way—I was operating on cervical and lumbar herniated discs, doing decompressive procedures, and fitting patients with halo braces.

Still, just as my interest in the spine mushroomed, as a result of doing those lackluster cases that no one else wanted, the issue of whose domain spine surgery should fall into—the neurosurgeon's or the orthopedic surgeon's—began to heat up in political and legal circles. I was still under the radar, and I couldn't foresee the trouble ahead, but the road I was traveling would, in the coming decade, open me up to hostile attacks on my professional integrity and the threat of legal prosecution. It would also reveal to me to my true vocation.

CHAPTER 11

A MEDICAL JUGGERNAUT

Big results require big ambitions.
HERACLITUS

ON A MILD clear evening in the spring of 1983, Lynne and I found our-selves members of a party for twelve at a resort called John Gardiner's Tennis Ranch, a desert retreat tucked into the lower slopes of the land-mark Camelback Mountain. Crisp linen tablecloths, shining, spotless wine glasses, polished wood and soft leather: the elements of fine dining glowed under wood beams and soft lights. A stunning view of russet crags—the "Praying Monk" side of the mountain—filled the wall-to-wall window on the west side, beyond which the sun dipped, buttering low clouds with a pearly pink. If you wanted to showcase the best the Valley of the Sun had to offer, the Tennis Ranch would easily make the short list of locations.

The dinner invitation had come from the big man at Barrow him-self, Dr. Green. It had struck me as mildly interesting when I received it. Apart from teaching the residents and picking up a few cases at Barrow, I was not associated with the founder and director. If anything, during my first couple of years I got the impression he disliked me: I was just another hungry neurosurgeon siphoning off the business. Still, though I was surprised to get the invitation, I didn't imagine anything was cook-ing as far as my future was concerned.

I figured that my presence and that of other colleagues from BNI was simply part of a process that had begun two years earlier when John Green began the search for his successor. Just the year before, that search had narrowed down to a young and charismatic vascular neu-rosurgeon named Robert Spetzler from Case Western. Dr. Spetzler had been out for one interview. Now he was back with his wife, Nancy, for a second look around.

The evening was pleasant but I still felt like a fish out of water. I knew the BNI surgeons just about as well as Spetzler did at that time, maybe less. Lynne and I mingled with the others over cocktails. I had a brief, unremarkable chat with Spetzler. We ate a highly curated and elegant meal. A small combo started up, and we watched as Dr. Green glided a slender, attractive woman with honey-colored hair and a wide Susan Hayward smile around the parquet dance floor. It was Robert Spetzler's wife Nancy.

"She's a beautiful woman," Lynne said." And what impeccable taste."

"Yes, very attractive," I said. "And Dr. Green is quite the hoofer, isn't he?"

We sipped our wine. Nancy Spetzler moved gracefully, elegant in a maroon suit with slim skirt and ruffled collar."

"He's quite handsome as well," Lynne said.

"Who? Dr. Green?"

"No," Lynne laughed. "Dr. Spetzler. You mentioned you'd met him before but you never told me he looked like a movie star."

"There was nothing to tell," I said. "It was during those rounds I attended at Chase Western. He was there but we didn't have any contact."

Lynne raised her eyebrows in a way she had that always preceded an astute insight. "Well, it looks like you may be having some contact with him in the future," she said.

The dinner was on a Saturday night. The next morning around eight o'clock the phone rang. It was Dr. Green.

"Hello Volker," he said. "Quite a dinner last night wasn't it."

"Yes," I said. "Lynne and I enjoyed it very much. Thank you for the invitation."

"Listen," said Dr. Green. "I know it's short notice, but Dr. Spetzler and his wife are flying back to Cleveland later today. I was wondering if you'd mind my bringing them by your house around 11:00, to see how you live over there in Arcadia. It's a nice area. I'm sure they'd appreciate it."

"Sure," I said, suppressing my surprise. "It's not a problem. Lynne and I will be happy to show them our home."

"Great. See you then." And he hung up.

Lynne had walked up while I was on the phone. She raised her eyebrows at me. "Did you just agree to something?" she said.

"Yes, it was John Green. He wants to bring the Spetzlers over to see the house before they head to the airport."

"To see our house?" Lynne said. "This morning? Why?"

"To get an idea of the neighborhood. I guess they're considering all the factors before they make a commitment."

Lynne's eye's widened. "Well, that's brunch time Volker. Are they expecting to be fed?"

"Oh, I don't know," I said. "But you're right. I guess we'd better prepare something."

Fortunately Lynne had done the shopping for the week. An hour later we surveyed the full German brunch we'd managed to pull together: different cheeses, meats, jams, boiled eggs, fruit and vegetables, smoked fish, and a variety of rolls and bread. The scent of freshly brewed coffee wafted from the kitchen.

With everything in place, Lynne went out to the front yard with the children. At four years old, Alissa was a little blond fairy, and our son Christopher, who had been born in 1980, a curly-haired cupid. Alissa was riding her tricycle in the driveway, with Chris standing on the step in the back holding on to her. Lynne, in jeans and a simple knit top, was watching them when Dr. Green and the Spetzlers drove up.

As it turned out, the brunch was unnecessary; the Spetzlers had already eaten breakfast, but we talked over coffee and got to know them. They'd checked out some of the nicer neighborhoods in Phoenix—Paradise Valley, downtown central Phoenix, around the Biltmore—but Nancy had not seen any houses she liked, and the neighborhoods were so starkly deserty she had a hard time imagining a move from cool, green Cleveland. They seemed charmed with the leafy, lawned Arcadia, though. More significantly, the scene outside had touched something in Nancy. The Spetzlers were starting their own family, and visiting us in our home away from the orchestrated events seemed to strike a chord in her. Later Nancy would tell Lynne that in those few moments outside, she decided that Lynne was going to be her new best friend.

The visit was short, but it was my first casual moment with Robert Spetzler. As I'd reminded Lynne at the dinner, I had met him briefly during the time I worked in Youngstown, Ohio. To stay current on neurosurgical advances, I'd attended rounds in Pittsburgh and had, on occasion, stopped in at Case Western in Cleveland too. Robert had been a junior attending at the time. But we'd never interacted.

Now he was sitting on my couch, long legs crossed, an easy, engaging smile, thick brown hair boyishly parted in the middle and still untouched by gray, as was his mustache. I sensed the athleticism in him, and later learned he'd been a competitive swimmer. He had fine hands, with long slender fingers befitting a top-rate surgeon, or a classical pianist.

I also resonated with his German origins, though unlike me he spoke English without a trace of an accent.

"So Robert," I said. "Your family is German?"

"Yes, he said. "I was born in Würzburg, Bavaria. I came over when I was nine."

"No kidding," I said. "I was born a hundred miles away, in Prussia. You got a head start on me though. I was twelve when my family came over. We must be about the same age?"

"I was born in November 1944," Robert said.

"Is that right? So was I."

"Where'd you settle in the States?" I asked.

"Illinois," he said. "You?"

"We came here, to Phoenix. Stayed in Arizona through medical school. Then I went to Boston. Did my training under Ben Stein at Tufts New England Medical Center."

Robert nodded, and Nancy spoke up. "Did you two meet in Boston?"

"Oh no," Lynne said. "Volker and I met during his internship year at University of Arizona Medical Center. I was finishing up my nursing studies, noting something on a chart in a patient's room when he walked in."

"Oh, you're a nurse?" Nancy said. "So am I."

"Well, I stopped working after Christopher was born, but I was a nurse practitioner until then."

"When did you two get married?" Nancy said.

"1974."

"Oh, that's too much!" Nancy exclaimed. Robert, can you believe it? We got married in 1974 too. April 17."

Lynne and I exchanged an amused glance. "We were just a week behind you," I said. "April 24."

The four of us sat sipping our coffee and shaking our heads over the coincidences. Dr. Green looked on, a satisfied smile on his avuncular, bespectacled face. He had fulfilled his goal, I think, in showing the Spetzlers a more intimate side of the community that—he hoped—Robert was on the brink of joining.

Robert and Nancy returned to Cleveland. Things went on as usual. I had no idea if Spetzler was close to making a decision or not, or if Dr. Green had other candidates for the directorship in mind as well. Either way, I didn't see how the outcome of the Spetzlers' visit could affect me.

I knew these things took time. Dr. Green had already gone down one dead end. In 1981 he had hired a neurosurgeon out of Harvard named Bob Crowell. Dr. Crowell was at Barrow a year, but it wasn't a good match, so a different tack was tried. Consultants were called in to quiz the doctors who worked there—including those like myself who did have some cases there but who were not on the teaching staff—to find out what qualities or skills these doctors felt a director should possess to make his leadership at Barrow successful. One of the suggestions that came up, more than once, was that whoever became the next director should include the major players or producers from the community—i.e., those neurosurgeons in town who were doing important cases—as part of the BNI team. A new search was initiated and about thirty candidates invited in for interviews. Despite my not being on the academic faculty at Barrow, and only on the periphery of this process, I was occasionally called in to take part in the interviews. But I had not been involved in Robert's.

A couple of months after the dinner and brunch, I returned to my office to find that Robert had left a message with my secretary. He wanted to talk with me. We arranged a meeting in an office at BNI. It was very brief.

"Good to see you again Dr. Sonntag," Robert said, shaking my hand.

"And you," I said. "How have you been since we met?"

"Fine, fine," he said. "So, how are things going for you? Do you like it here?"

I wasn't sure if he meant Phoenix or Barrow, but I responded in the same general vein.

"Things are going well," I said. "You know how it is to build a practice. But I'm getting more patients now. Things are good."

I got a sense of something going on, but Robert was holding his cards close to his chest. We parted cordially, and I forgot about it.

Until a month later. Robert called me again. Again I walked over to Barrow and found him in an office. This time he was a little more forthcoming.

"Thanks for agreeing to meet with me again. As you know, I am considering coming out here."

"Yes, I am aware of that. It's a big decision," I said.

Robert pulled on his mustache. He nodded. "Yes, there are a lot of things to consider."

A moment passed. We smiled at each other. Then....

"What do you think about forming a group at Barrow?" Robert said.

I wasn't quite sure what he meant. Was he asking me if I wanted to be in a group? Was it just a general question?

"It might be time," I said. "BNI has doubled in size since it opened. It's acquired a solid national reputation. More cases are coming in."

Robert nodded again. "Of course I haven't made up my mind, but it is a consideration."

He stood up, thanked me, and said he would be in touch.

I walked back to my office hardly aware of the cars passing in the street or the sun beating down. Something was cooking. I needed to run it by someone I liked and respected. I called Jack Kelly.

Jack Kelly was an amiable, gentle man in his early fifties. He was a good neurosurgeon, his patients loved him, and the two of us got on famously. Short and stocky, he had a big head of gray hair. He also had a rascally sense of humor.

Jack was the neurosurgeon I had come to an arrangement with in the beginning days of my practice; we covered for each other on alternating weekends so we could each get some time off. Every Friday afternoon, whichever one was off would give the other the lowdown on which patients he had and where they were. Jack practiced out of Baptist Hospital, but his patients, like mine, might be anywhere, so when I was on call, I'd spend the weekend running around to check up on all those folks.

One Friday, Jack included in his list a Mr. Jones down at Good Sam. It had been a complicated case, an aneurysm in the anterior communicating artery. The anterior communicating artery is the blood vessel of the brain that connects the left and right anterior cerebral arteries; it's part of the cerebral arterial circle known as the Circle of Willis, the freeway loop of the brain with exits branching off to other important arteries. One complication in surgery on this part of the brain is a cerebral occlusion in the anterior communicating artery, which can lead to weakness or even complete loss of leg movement.

I went on my rounds that weekend and eventually came to Mr. Jones. He was alert. He seemed to be recovering well. But I needed to check a particular response to make sure he hadn't suffered a stroke, which would cause that weakness in the legs.

"OK Mr. Jones," I said, "move your legs."

Mr. Jones gave me a quizzical look.

"Move your legs," I said again.

I looked down at the sheet. Not a twitch.

"Can you move your legs Mr. Jones?" I said.

Again nothing, except an increasingly disgruntled look from Mr. Jones.

Shit, I thought. *The guy's stroked out.* I tried one last time. "Lift your legs."

It was clear nothing was going to happen. Disheartened, I lifted the sheet to uncover Mr. Jones's legs. Jack Kelly's face flashed in my mind. *You son-of-a-bitch*, I thought, shaking my head. Mr. Jones was a double amputee; he had no legs. To this day I have no idea why Mr. Jones didn't just say so, or the nurses, for that matter, who stood there watching me make an ass out of myself. Maybe they were all too embarrassed.

That was Jack Kelly: decent, human, all there for the patients, and not above a good joke. Now, getting a whiff of whatever it was that was cooking at Barrow, I wanted to be completely transparent with him. I got him on the phone. His response didn't disappoint or surprise me.

"Jack," I said, "I want to give you a heads up. Something's going on at Barrow. This guy Spetzler asked to meet with me. It might be a good opportunity. Can't tell yet, but I might have to beg off on taking calls at some point here."

"Of course," he said. "Do what you think is right." I knew he meant it.

Things clicked into a higher gear after that. Spetzler kept coming out to Phoenix. He and Dr. Green arranged larger meetings that included neurosurgeons already on the staff at Barrow, as well as those like me who were operating there on a regular basis. There were eight of us in all: John Green, Robert Spetzler, Tim Harrington, Phil Carter, Hal Pittman, Andy Shetter, and myself. Carter and Harrington had their own group that included a surgeon named Bill White, and it was eventually decided that he should also be invited. This is not to suggest that Spetzler was any more committed outwardly to the new venture than he had been so far. "Let's have another talk about preliminary possibilities," he'd say.

The cat-and-mouse game had commenced. For the next six months, we met every other week at a large, oblong table at the Plaza Club on the top floor of the Plaza Tower. As the twinkling lights of the city stretched out to the horizon on all sides, we fortified ourselves with dinner and went to it. Each meeting ended in the same deadlock. Robert would say he might come out if we agreed to form a group; the rest of us agreed that we would form a group if he came out. Nobody wanted to drop the shoe. Sometimes I came away from those meetings thinking, *what the heck went on there?* The next morning I'd tell Lynne, "I don't know if this thing's going to go anywhere. I just have to get back to work." I was keeping Jack Kelly up to speed on the process, and he continued to be supportive. "If you think it's right for your career," he said, "go for it." The problem was that for months, there was no solid "it" to go for.

During this initial period of negotiations, a third component arose that illustrated just how tedious and crazy the whole process of forming a group was going to be. The prospective group members wanted an assurance from St. Joe's that if we formed a group, we would have offices on the premises of the hospital. St. Joe's did not want to commit to that until they had an assurance that we were indeed going to form a group, and we weren't going to form a group until we knew Robert was coming out, but he wouldn't come out until we formed a group. It went round and round like this, with no one willing to set his domino on the board first.

The reluctance to make the first move is understandable. There was a lot at stake for all the players at the table—a lot of money—and there was a dizzying set of issues to be worked out before anything could move ahead. To add to the uncertainty, the eight of us didn't know each other well, and we knew Spetzler even less. Eventually, lawyers and advisors got involved, and over the course of many weeks, night after night, till one or two o'clock in the morning, we parked ourselves around the table and hammered it all out. We drew up the bylaws, agreed upon a name—Barrow Neurosurgical Associates—and formulated the rules and regulations for the group. Details had to be addressed: as far as call schedules, pensions, disability and how we were going to get paid; as far as how many times we could go to a meeting and what our initial pay for a meeting should be; as far as what should be considered shared expenses (malpractice insurance, rent, and overhead) and what should be unshared expenses (trips, for example). All the possible points of contention in a financial relationship had to be laid out and dissected, like the various parts of a cadaver on an examining table.

The process moved faster after that. St. Joe's agreed to provide us a facility on the site of the old emergency room. While that was going up, the eight group members tackled the task of setting up a financial entity into which we could merge the business end of our practices. That was another complicated business. Between the eight of us, there were four cost centers: one for my practice, one for Spetzler, one for the group already at Barrow, and one for the outside group. How could these four cost centers be factored in when the new group formed? How would we pay the initial bills that came in? After all, it was not like once we started doing cases the money would roll in immediately. It takes about six months between the time you do an operation and the time the insurance money comes into a physician's account. The only way to manage that crossover period was to borrow money from the bank.

In the end, we decided to ask Valley Bank for a loan of two million dollars. That is admittedly an intimidating amount of money, and one of the physician's wives initially balked. She did not want to dive off that board, and it took her a while to agree to sign it. When, with the rest of us, she took the plunge, we got the money, formed a corporation, and in October 1983 it all came together. We were in uncharted waters. We had no idea if we were going to be successful. We were all putting our trust in other individuals that we didn't know well. But, we were open for business. It was a go.

CHAPTER 12

THE COMPLETE SPINE DOCTOR

Let me tell you the secret that has led me to my goal.
My strength lies solely in my tenacity.
LOUIS PASTEUR

BY NOON on a typically scorching day in June of 1987, a smattering of bare-chested youth in cut-off blue jeans had gathered at the Blue Point Ridge on the Salt River northeast of Phoenix. A place where great jagged rocks rise in stark cliffs from the river, it is still a favorite destination when temperatures reach the triple digits.

I remembered coming to places like this as a boy, a guest—with my parents and younger brother—of George Beuchel, the sponsor who had helped my family immigrate to the US. We'd picnic and play cards on the treeless and rocky shores, frying our pale German complexions in the intense sunlight. We'd float down the river on the inflated inner tubes of giant tires. My father enjoyed those outings. Wading into the wild river and letting the rushing current catch him, he became his old adventurous self. Later, as the shade from the cottonwoods and mesquite trees deepened, and great cumulous clouds rose behind the low brush-covered mountains, he'd sit back, crack open a beer, and get a short reprieve from the depressing reality of his circumstances.

So much else had changed in my life, but these surroundings were the same.

Today it was not just the usual crowd of tubing enthusiasts. A contingent from Barrow was there as part of an initiative I'd organized in my new outreach role; the campaign was the offshoot of a national program called Feet First, First Time aimed at educating swimmers on safety practices. It's main message was that you needed to take precaution before diving into some pool—natural or manmade—that you might not have used before, or might not have visited for some time. Just because the water was at a depth of eight or ten feet the last time you jumped off a cliff into the river didn't mean that the water level hadn't changed and that the depth was now only four or five feet. This was especially important in a place like Arizona, where you can get drastic drops in the water level from one year to the next, or even from one week to the next. You still might break your ankles if you jump feet first into shallow water, but that is far preferable to breaking your neck or fracturing your skull.

We'd organized the event at the Salt River Cliffs because it was the site of numerous accidents. We had the media come out, we invited former patients with spinal cord injuries—one, strapped into his wheel chair with his atrophied arms resting in padded armrests, still wore a neck collar—and we talked about how to prevent such injuries. Now amid yellow banners and stacks of matching can coolers proclaiming FEET FIRST, FIRST TIME, I hoped the life-saving message I was there to deliver to the largely young, male gathering would penetrate their skulls before their encounter with the riverbed either killed them outright or turned them into quadriplegics.

In the four years since I'd become a full member of the staff at Barrow, I had built up considerable expertise in the spine. I owed the direction of my career—in some part—to all those trauma cases the residents sent me, which presented me with a whole gamut of spine injuries. Come summer each year, a number of fresh young men, their futures yawning wide and promising before them, became sadly similar cases in my career trajectory. They did this by launching themselves head first into shallow swimming pools or rivers with notoriously shifting water levels.

Such a launch was what indirectly brought one of the wheelchair-bound men to the river that day, a boyish, brown-haired man in his thirties named Andrew Dillon, who had already spent a decade living with the consequences of his rash impulse. To a neurosurgeon or an orthopedic surgeon, the possibility of injury would be foremost in your mind at a place like this—once you see the kind of damage that can happen in a split second of mindlessness, you never forget it. But most people don't think much about, and young people think about it even less. So Andrew was here to share his story, to paint a picture in the minds of the young on the devastating reality of a spine injury.

"When you get to the aspects of paralysis for the rest of your life," Andrew said to the reporter, "and how it affects not only you but your family and friends and those people around you, they just don't have any idea what it's like."

Like so many accidents of this nature, Andrew's might never have happened. Small decisions at several points along the way determined his fate.

Andrew's friends were going tubing on the Verde River that Sunday, another popular locale north of the Valley that offered rushing chutes interspersed with placid, floating pools. Initially, Andrew declined the invitation in favor of staying home and watching the US Open Golf Tournament on TV. But the tubes were in the trunk of his car, and since he felt obliged to deliver them to his friends anyway, he decided to go along.

It had probably been one of those idyllic summer afternoons. The young men floated in their fat, black donuts, toasting themselves and watching the thick brush pass by along the shore. Hours later, the shadows lengthening, the cicadas gearing up into a shrill racket, the group trudged back upriver to where they'd left the car. They packed up. A moment later and they'd have driven away. But one of the boys had an impulse for a last bit of horseplay. He threw a Styrofoam ice chest in the river.

Andrew probably replayed the next moment hundreds of times over the following decade. Without a thought, he dived in to retrieve the ice chest. His head hit a rock straight on, compressing the vertebrae in his neck. He felt immediate numbness. When he surfaced, face down and bloody, his friends pulled him from the river and laid him on a bed of inner tubes. But the damage had been done.

The spinal cord, housed inside the spinal column, is a delicate organ the thickness of your ring finger that controls all the body's functions under commands from the brain. Each level of the spinal cord controls a different function of the body. The cervical levels labeled C5 through C7 control arm movement. C7 specifically controls extensions of the arm. C6 controls the muscles that bring the arms inward, permitting such movements as feeding and back scratching. When one level of the cord is severely damaged, all sensation and movement below that point are imperiled. The cervical vertebrae just below the skull are critical to life itself; they control the respiratory system.

The spinal cord can be damaged two ways. First, there is the impact itself, which flexes, compresses, stretches or rotates—or even crushes or dissects—the spinal cord. After the initial assault, metabolic changes take place in the cord that can lead to further degeneration. Nerves gasp for oxygen. Chemicals are released that further constrict blood vessels, choking off the oxygen supply. At this stage the neck is unstable. The slightest movement or mishandling of the injury can prove fatal or cause further damage.

Seven cervical vertebrae, those closest to the skull, constitute the skeleton of the neck. Andrew's accident crushed the fourth vertebrae, labeled C4, and damaged the nerve roots in C3 and C5, leaving him paralyzed from his neck down.

As I listened to other spinal injury patients address the crowd from their wheelchairs, I reflected sadly on the fact that my message might save some, but it would fail to touch a great many other young men. I wanted to grab each and every one of those kids crawling along the shore, already jumping from the cliffs, and break their illusions of invulnerability. Their movement, independence—everything was at stake. "If it's a really bad break," I told the news crew that day "and they can't move their arms and legs, and they can't breathe, they could be on a respirator for the rest of their lives."

But I knew the percentages. In the two years preceding the FFFT event, ten people had broken their necks at the Salt River alone, all of them young, all of them male. Diving accidents were the fourth leading cause of paralysis in the county. Already spinal injury cases such as theirs increasingly took all my attention.

Whether it was Robert's and Dr. Green's leadership or the integrity, professionalism, and talent of those involved, the eight of us who hitched our stars to the new group wagon at BNI got along quite well. That is not a small feat when you had individuals who were previously calling all the shots for themselves, or who had been in a smaller group and knew their partners well. But we were all committed to building up the group and the institute, and willing to work hard at it.

Slowly our volume increased and the residency picked up, but it didn't happen overnight. It would take ten years for us to build up enough volume to work exclusively out of St. Joe's. Even when the group garnered some recognition and Barrow's reputation grew, we were afraid that if we stopped taking calls at other hospitals, we would interrupt our flow pattern, so we kept taking cases all over town—each associate continuing to cover the hospitals he had served—well beyond the time it was reasonable to stop. It took a long time to work all the angles out.

Some of those angles added a considerable load to already crowded schedules. Take calls, for example. While all eight of the associates had taken calls at some hospitals, now whoever was on call over a weekend had to cover all eight or ten hospitals that the associates combined had served. Certainly it helped to have eight doctors divvying up the load—which meant that no single doctor was on call as frequently as before—but when your turn hit, it was a real monster show.

Every Friday afternoon the group would get five or six or eight pages of patients currently under care, categorized by hospital. Whoever's turn it was—and that included Robert Spetzler—would easily spend ten hours on Saturday crisscrossing the valley to make rounds on them: up to John C. Lincoln in the north, Good Sam down by the airport, Phoenix Memorial Hospital and St. Joe's in central Phoenix. It wasn't just the distance either. You might have privileges at some of those hospitals but not at others. If you didn't have them yet, you had to get them, not only to take calls but in case an emergency surgery arose.

Calls may sound easy, but they weren't. They were painful. They robbed you of your peace of mind. They could interrupt anything and they usually did. You might have dared to go out to dinner and your beeper would ring. You had to get up, go in search of a phone, and ring in to the answering service. If you were lucky it might just be a simple prescription renewal. But even then, by the time you talked to the patient—"Doctor, I'm Mrs. Jones, a patient of Dr. Harrington's, and I need my Percocet"—called in the prescription and returned to your table, your food was cold and your companions were starting on coffee. Forget drinking. You might order one beer, but that's all you could risk. And you could never go anywhere far from a telephone; you were tethered to the phone. Even today I keep a pair of folded scrubs by a chair next to my bed, a holdover from those midnight calls.

Calls were lucrative though, when the patient got to you and when he had insurance. The first thing you had to do was determine if, based on what the resident told you when he called on a weekend, you wanted to take the case. "We've got a guy here with severe back pain," the resident might say. "I don't think we should discharge him. Let's admit him." "Okay, yes, I'll see him tomorrow on my rounds," I'd tell him. So then I'd see the patient during rounds. We'd do a diagnostic study—let's say the patient had a big, ruptured disc—and I'd operate on Monday or Tuesday.

Fifty to sixty percent of the time the patient had insurance—on a good weekend that number might hit eighty percent. But it was more usual to see lower; forty percent of the time you were on call, you weren't going to get paid for the subsequent cases. Still, if the residents had thirty admissions, and you later operated on four of them, you'd get paid for two or three of those surgeries, and right there you'd make maybe $7,000. Those would include all kinds of serious cases: head trauma; skull fractures; blood on the brain; epidural, subdural, and intercerebral blood clots; spinal fractures; lots of emergencies. Interspersed with those surgeries, you might make a hundred dollars doing a consult. Even if was only taking a history and physical—H & P—you could, and did, charge for that.

That may sound like a tidy sum just from the cases that came in on call, but all told, we worked hard for what we made. The weekends on call easily pushed our total time for the week up to 120 hours: eight to ten hours just scrambling from hospital to hospital, then making rounds, doing an exam here, a four- or five-hour operation there, back in the car to see a patient that you hadn't called in on yet, checking on your colleagues' post-op patients—all without neglecting your own patients in the process.

Looking back, I honestly don't know how I did it. And those were only the cases from the weekend emergency room. Gone were the days of twirling my thumbs in an empty office waiting for the phone to ring. Gone were the days of simply practicing neuro-

Sitting at far left next to the founder of BNI, Dr. John Green, shortly after the formation of the Barrow Neurosurgical Associates group in 1983. Five years earlier he had told me I would starve if I wanted to practice neurosurgery in Arizona. New director, Dr. Robert Spetzler, is at the far right. Photograph used with the permission of Barrow Neurological Institute.

surgery, as demanding as that was. The academic side of my career was about to take off too.

Judging from the hairs on the back of his neck, the pilot of the small four-seater plane seemed totally unconcerned at the sudden plunge. One second we were executing a slow ascent to 20,000 feet and the next the floor of the plane lurched out from underneath us. I think I screamed. Who could tell? The cramped cabin reverberated with the din of the unmuffled engines and the screaming wind. My entire body went rigid in a futile attempt to hang on to something, if only myself. My eyes darted to the window. The urban grid visible beneath the wing slapped up under us while the clouds that had been gently floating by whooshed up as if pulled by a trip wire.

"Gave you a start, didn't it?" the pilot hollered over the racket.

"Scared the beejeesus out of me," I threw back.

"It's just these small planes," he said. "They hit the heat pockets in summer and just kick out for a minute. It can rattle you the first couple of times, but you get used to it."

I don't think I ever got used to it, but in the first couple years as an associate at Barrow, flying up north to Prescott or Kingman, or out west to Benson or Havasu City became a regular part of my duties. There was no other way to conduct another new outreach program—one through which we provided updates and talks to the surgical staff in small towns around the state—but to board one of those aerial go-carts. We only did it in the summer, which meant we often hit those heat pockets. I never felt secure in those little planes, and when they hit one and my seat suddenly dropped out from under me, I'd start praying.

That afternoon we were headed to Winslow, a former railroad town on the old Route 66 (now Interstate 40) halfway between Flagstaff and Gallup, New Mexico. As we climbed again, the air came back into my lungs and I relaxed. The sky and land resumed their natural horizontal relationship, and I went back to surveying the scene below. It would be an hour before we reached Winslow. Might as well enjoy the view.

Below me the mountain-ringed bowl of Phoenix opened up. The city had grown tremendously since I'd stepped off a train in downtown Phoenix in 1957, just a few days off the trans-Atlantic refugee boat and still trying to digest the vastness of the country I had traveled through on the amazing Super Chief. I'd changed too, from a green immigrant kid who'd been astonished to find that oranges grew on trees to an up-and-coming spinal neurosurgeon.

The plane touched down and my thoughts returned to earth with it. Here I was in a remote town best known for its reference in an

Eagles song. The sun was going down—my visit always coincided with the monthly doctor's meetings that took place in the evening. I'd be speaking to generalists, not neurosurgeons, and those doctors who did show up were often there primarily to earn continuing medical education credits. The last time I'd done the outreach, only three doctors showed up. I doubted Barrow would get any referrals out it, which was, in truth, the underlying motive for the endeavor. I stepped out of the plane. "Catch you later," I called to the pilot. "With any luck, we can get home before ten."

During the few years that I did the FFFT and outreach programs, neurosurgery was becoming more complex and specialized. In the mid to late 1980s, my associates and I were still doing all areas of neurosurgery, but as our careers moved on and neurosurgery changed, we started subspecializing. We had seen this coming, and one of our intentions in forming the group in 1983 was subspecialization. We wanted to know what could be done that would attract not only patients but also referrals from other physicians and especially from other neurosurgeons. It was great foresight on Robert Spetzler's part to encourage us in this direction.

I was still doing cranial cases, but as my reputation as a spine doctor grew, I got overwhelmed with spine cases to the point that I didn't feel comfortable doing cranial cases anymore. Surgery is like classical piano; repetition is key to developing skill. So, by the late 1980s, I was easing off on cranial cases and concentrating on the spine.

Such a move toward specialization was true of my associates, too. By then, Robert had stopped doing spinal cases. Another doctor, Andy Shetter, had gone the pain and epilepsy route. If we didn't have someone on the team who specialized in an area of need, we hired someone new. Pediatric neurosurgeon Hal Rekate had joined us as early as 1984 and later worked on the T. J. Case with me. In that way, we all started niching out.

It wasn't just cases involving injuries to the cervical spine that I was doing either, but surgeries on all parts of the spine from the occiput down through the long thoracic curvature, the lumbar curvature in the small of the back, and the sacral curvature that ends at the tailbone. I was taking out ruptured disks and tumors wherever they occurred along the entire spinal cord. Spine surgery was not considered outside the domain of neurosurgery, and though admittedly it was more common for an orthopedic surgeon to tackle problems in the lumbar region of the spine and a neurosurgeon to do the cervical, now both orthopedic surgeons and neurosurgeons took on these kinds of cases. The only thing a neurosurgeon could not do that an orthopedic surgeon could was to instrument the spine. And because instrumentation and fusion procedures were just breaking on the horizon for neurosurgery in the early 1980s, no one had really challenged that convention.

I thought about the direction my career was taking on that plane. I enjoyed teaching the residents and doing the outreach. But the teaching was still heavily weighted towards the brain. If my own trajectory had naturally taken me more deeply into the spine, I thought, other neurosurgeons must be moving in that direction as well. Judging from my experience, it seemed to me that the spine would almost certainly be a new specialty area for residents. An idea began to formulate in my mind. Could I build a more formal course of study on the spine? Might I begin offering a post-residency training? In short, could I start a spine fellowship at Barrow?

Little did I know that these very questions would soon embroil me in a battle over one of the most controversial neurosurgical topics of the decade—neurosurgery's role in the spine at a time of game-changing technical advances.

CHAPTER 13

TURF WARS

A merely fallen enemy may rise again,
but the reconciled one is truly vanquished.
FRIEDRICH SCHILLER

ON A LATE summer evening in 1990, I sat at a table with three orthopedic surgeons, my fellow neurosurgeon, Richard Douglas, and three or four administrators from Saint Joe's. The only light in the room came from an X-ray box on the table. Richard and I were present because the executive committee of the hospital, under the influence of the orthopedic surgeons, had demanded that I submit to the orthopedic committee for review my reports on three spine cases I had done. The bone doctors had duly gone through my reports and were now here to "evaluate" the surgeries.

Since I had successfully made my case for a neurosurgical spine fellowship at Barrow two years earlier, there had been periodic eruptions of alarm from the orthopedic camp. That was in 1988, the same year the orthopedic surgeon, Dr. S., walked out on me during that first spine surgery using the lateral mass plates. Then came TJ in 1989, a resounding vindication of the use—and the invention—of alternatives to wiring the spine—especially on the part of neurosurgeons. But now I'd rankled the orthopedic surgeons even more by intruding into an area they considered their inviolable domain—instrumentation of the thoracic-lumbar spine. So I was hardly surprised when I received the summons to what I knew would be a kangaroo court.

Richard Douglas knew it too. Richard had been my third fellow and had only recently finished the spine fellowship when the meeting was called. Though he'd joined a private practice with a local neurosurgeon, he had privileges at Saint Joe's, and he'd assisted me on the three spine cases now up for review. He wasn't a senior surgeon, but the prospect of a dogfight didn't intimidate him. Richard was a brash, funny, rough-and-tough kind of guy, a former marine with a marine attitude about everything, including women. If you met him on the street, the last thing you'd picture him doing was handling delicate brain or spinal material.

It was after 6:00 p.m. when Richard and I arrived in Conference Room B. The orthopods and administrators sat on the far side of the table, their faces barely visible in the gloom. The X-ray box glowed blue to one side. Richard and I took our seats across from them. Directly across from us, in the center, sat the chief of staff, John Bull, and to his side a combative, in-your-face orthopedic surgeon named Allen Zimtbaum. Even in the dark Zimtbaum's tall, sturdy form was formidable. Light glinted in his abundant, wavy hair and reflected off his thick glasses. For Allen, all issues were black or white, never gray. The others were well-established orthopedic surgeons with privileges at St. Joe's. I had done rounds with all of them, had partnered on numerous surgeries with them, had shared a collegial joke or two. Now I had the sense of dogs, teeth bared, straining against their leashes.

Dr. Bull opened the proceedings with a loud clearing of his throat:

"As we are aware, Dr. Sonntag is starting to do thoracic-lumbar instrumentation. We're here today because the orthopedic community has asked to review several of his recent cases. Coming from experienced, established spine surgeons, this request seems appropriate. I'm sure we all agree that our charge is to ensure we are not harming patients. All right. I'll turn it over to Dr. Zimtbaum then."

A short pile of films rested next to the viewing box. Zimtbaum nodded at them. "Please clip that first X-ray up on the screen, Dr. Sonntag," he said, in a guttural bray. "And tell us how you diagnosed it."

I glanced at the X-ray. The thoracic spine followed its natural curvature until it reached the junction with the lumbar spine; there it had fractured, causing the vertebral column below to shift, like a row of books pushed out of line. "It's a fracture dislocation of the thoracic lumbar junction, a three-column injury, very unstable."

"How did you arrive at your assessment of the instability?"

"It's obvious on the X-ray," I said. "I'm familiar with Denis's three-column model. The break here clearly involves the posterior, middle, and anterior columns of the spine."

Dr. Zimtbaum inhaled nasally, expelling the air in a burst.

"Tell the review committee Dr. Sonntag, if biomechanics was part of your training."

"No, it wasn't," I said. "But as you might be aware, I recently took Dr. Sandy Larson's intensive course on thoracic lumbar instrumentation and fusion. Dr. Larson has been successfully performing complex spinal surgeries, including instrumentation, for over a decade."

"And Dr. Larson is a neurosurgeon?"

"That's correct."

"So you haven't been trained properly."

"Only if you insist on not recognizing Dr. Larson's well-respected work as training. Moreover, I believe I've worked with all the orthopedic surgeons here on decompressions, staying to observe the instrumentation. Once I'd done the formal training with Dr. Larson, it was a natural step to begin doing it myself."

"I see," Dr. Zimtbaum said sourly. "And you consider observation equal to training." He paused to let his colleagues appreciate his point. "Please show us the postop slide now."

I replaced the slide with one showing the double rail of instrumentation.

"Please explain the procedure you did on this injury."

"Given the severity of the case I thought it was appropriate to stabilize the spine using Harrington rods and fusion, the latter of which I've successfully performed numerous times on the cervical spine, and both of which I've observed repeatedly in partnering with the orthopedic surgeons on thoracolumbar spine injury cases."

Dr. Zimtbaum murmured something to the orthopod next to him. There was a brief consultation and a nodding of heads. Then Zimtbaum resumed.

"Right," he said. "Fusion. We have some serious concerns about your understanding of fusion Dr. Sonntag. As with instrumentation, it was not part of your neurosurgical training."

"That is correct. But it was covered intensively in Dr. Larson's course, the entire focus of which was instrumentation and fusion of the thoracolumbar spine."

"At which levels did you fuse the spine in this case?"

"The fracture was at T12. I fused two levels above to T10 and two below to L2."

"Didn't you think it might be better to fuse three above and three below, given the severity of the fracture? What was your rationale for only doing two above and two below?"

"I felt I had attained a good enough purchase with two above and two below. I didn't see the need to do three levels."

"You would agree though that with an injury like this most orthopedic surgeons would fuse three levels above and three below."

"Based on the latest literature and my experience, and given the pathology, I thought what I did was the appropriate thing to do," I said.

A short break ensued while the orthopods whispered and nodded. One of them rose, obviously agitated. I heard low snatches: *fusing around the spine…doesn't understand the biomechanics…yoyo…waste of time.* Then he walked out.

The meeting resumed but I agreed with the dissenter. It was a waste of time. A hostile stink of pre-formed conclusions hung in the air. Nothing I said was going to make any difference.

And so it went with the other two cases. I didn't understand the biomechanics of the spine; I didn't understand how the spine moved under normal functioning; I wasn't equipped to determine the proper instrumentation to restore what had been lost to injury; I was endangering patients when I did sublaminar wiring—I hadn't been trained, they continued to insist. It made no difference to them that the outcomes on all three surgeries had been good. No infection. No neurological deficit. When, after two and a half hours, the meeting broke up, a sour taste filled my mouth.

"Well that was a real butchering," I said to Richard once we'd rounded the corner into a short convoluted hallway. It was close to 9:00 p.m. We passed an oversized portrait of Dr. Green, paternal and beneficent. I wondered what battles he'd fought over the years.

"They were never going to accept any justification," I said.

"Of course not," Richard said. "They were speaking for the whole orthopedic community. They had to say you did a sorry job. The meeting was just a formality to get it on record."

"And those questions on my infection rate. I think I've had one infection out of the ten or eleven thoracic-lumbar cases I've done. They know the infection rate goes up whenever a foreign body is introduced. Instrumentation carries that risk with it."

"It didn't have anything to do with the cases under review anyway," Richard added. "The infection rate was a side swipe."

We pushed through a plain door into the BNA offices, passed the empty examining rooms, and stood for a moment before parting. I was still fuming. "We operate on patients with all kinds of complications and behaviors, people with diabetes, people with bad wound care. They have a low incision and they wipe their butts and the next thing you know they've got an infection. There are things you can't help."

"It wasn't the infection rate," Richard said again. "Don't you think the real issue here, Chief, is money? That is, doing the thoracolumbar spine, we're stealing food out of their mouths?"

"That may be part of it," I agreed. "Thanks for your help Richard. You going home now?"

Richard gave me a devilish smile. "No, stopping in to visit a little nurse I met over at Scottsdale Memorial. Have dinner, a couple of drinks. Then I'm going to turn her upside down and drink her like a water fountain."

I shook my head. "You're something else Richard," I said. "See you around."

Richard's audacity had lifted the stone off my shoulders for a moment. But as I gathered my papers and slipped them into my briefcase, it finally struck me: What I was feeling wasn't anger; it was betrayal.

The conflict over instrumentation of the spine had been brewing at least since I established the neurosurgical spine fellowship in 1988. A fellowship is an intense period of training after the residency in a specialized field. Usually lasting a year, it gives physicians the opportunity to intensify their learning in a subspecialty field, in this case, spinal neurosurgery. Except for Sandy Larson, who had been the sole champion of the neurosurgical treatment of complex spinal cases, including instrumentation, since the early 1980s, neurosurgical spine fellowships were unheard of. As far as I knew, the one I had established at Barrow was one of very few in the country.

Organized neurosurgery was jolted out of its complacency regarding the spine, though, when we found out that the orthopedic community had convinced an important licensing body that governed what residents and fellows should be trained in—the Accreditation Council of Graduate Medical Education, or ACGME—to begin formal consideration of spine fellowship training under the sole auspices of orthopedic surgery. That meant that spinal instrumentation would be taught to orthopedic residents and not neurosurgical residents. And since instrumentation is a vital part of spine surgery, once orthopedic fellowships were accredited, the orthopedic community would take over complete care of the spine.

Such a move would hit me very hard, I knew. If the orthopedic community succeeded in having neurosurgeons barred from doing these procedures, much of the expertise I had gained in the area of the spine would be wasted. Not only would I not be able to take cases that I had been steadily increasing my knowledge of and gaining experience in over the last decade, I wouldn't be able to teach residents what I had learned. I would have to satisfy myself with simple spine procedures or "return to go" and start back up again with cranial cases.

The howl that went up in the halls of neurosurgical offices and at the national meetings could be heard on the moon. When the news circulated at the association meetings—the American Association of Neurological Surgeons (AANS) and the Congress of Neurosurgical Surgeons (CNS)—panic broke out: *"We're going to lose the spine; the orthopods'll take instrumentation; they'll have an accredited fellowship in the spine; where will that leave us? We've got to do instrumentation."*

In face of this imminent threat, organized neurosurgery convened a Spine Task Force to define the neurosurgical role in the spine and the role of the spine in neurosurgery. This spawned numerous subcommittees and resulted in the dissemination of guidelines on spinal training of neurosurgical residents and fellows. Most importantly, the Task Force stipulated that neurosurgeons should do instrumentation of the whole spine, not only the cervical spine, and that meant thoracic, lumbar, and sacral instrumentation. As vice-chairman of the task force, it fell to me to align this decision with resident and fellow training.

Being one of just a few neurosurgeons doing complex spine cases, in particular cervical operations, I also happened to be the secretary of The Joint Section on Spinal Disorders and Peripheral Nerves. I sent off letters to the multiple societies and organizations that governed neurosurgical practice and academics. I also sent a letter off to the Orthopedic Resident and Review Committee, stating the case for training neurosurgery residents in spine instrumentation. They didn't bother to respond directly to me. In fact, I think I provided them with some ammunition because they sent off a letter to the ACGME claiming, "the addition of spine instrumentation to a neurosurgical residency would NOT provide adequate training and would adversely affect patient care."

That was it. The battle lines were drawn. Nationally it would be a long-drawn-out war but within Barrow we could wage our own campaign. That's when, in July of 1990, I flew to Milwaukee and took Sandy Larson's intensive course, alongside my first three fellows: Steve Papadopoulos, Ian Kalfas, and Richard Douglas.

The main topic of Sandy Larson's course was precisely the procedure the orthopedic surgeons objected to my doing: instrumentation and fusion of the thoracic lumbar spine. Lateral mass plates had been introduced by then—the kind I had used on the surgery with Dr. S.—but they were used primarily on the cervical spine. For injuries farther down, surgeons (primarily orthopedic) depended on a complex system of rods, hooks, and wiring, the most prominent among them, the Harrington System and the Luque System, as well as a deviously complicated system developed by French surgeons, the Cotrel-Dubousset Instrumentation (CDI).

The Harrington System had been devised in the late 1950s by Paul Harrington, an orthopedic surgeon, to straighten the curved spines of scoliosis patients. It consisted of a stainless steel rod attached to the spine with hooks inserted into the vertebra at the top of the curve and the vertebra at the bottom of the curve. Once fixed to the affected stretch of vertebral column, attached ratchets were tightened to straighten the spine. A modification of that came in 1976 with a device created by a Mexico City physician named Eduardo Luque. Luque rods consisted of L-shaped, cylindrical rods but they were affixed to the spine by using 16- or 18-gauge sublaminar wires at multiple segments along their length. Then, around the time BNA was formed, Cotrel and Dubousset developed the CDI, which, with its many parts, was perhaps the most time-consuming and difficult system of them all. In the 1990s these devices, though soon to be superseded, were being used for other kinds of spine surgeries as well. In Milwaukee I found myself struggling to master all of them.

Richard Douglas partnered with me throughout the two-week course. That first morning I stood across from him at the stainless steel dissecting table in the stark, cold cadaver lab. On the table lay a black body bag.

"You do the honors," I said. "Let's meet our donor."

Richard unzipped the bag, revealing the waxy-looking corpse of a bald, elderly man covered with a sheet. "Hello Grandpa," Richard said. "Thanks for coming to the party."

We removed the bag and sheet and flipped the dead weight of the body over onto its stomach. I remembered how as a med student, I'd met my first body this way. From the back, it was much less personal. Richard and I dug in, dissecting the muscle from the spine for the next two hours until we had a clean column. As we pried the muscle from the bone—it was like chicken breast left too long in the oven—the pungent scent of formaldehyde and disinfectant seeped through our masks. I knew the smell would stay with me for hours, no matter how much I snorted it away.

The instrumentation did not come naturally; my first impression of the collection of rods and hooks was of tinker toy assembly. "Rats," I said, when, the second week, a hook I had carefully placed popped off. Richard and I were working with the CDI, Richard inserting a rod on one side of the spine and me mirroring that on the other side. At first we'd tried doing it simultaneously, but we'd soon learned that it was one side at a time—without wires, it was almost impossible to keep the rod stable while affixing the hooks. I'd gotten the hook firmly attached at the top vertebra, had inserted the rod while Richard stabilized the spine, and was carefully placing the bottom hook when the rod cantilevered out, sending the top hook flying.

"This thing's impossible," I said, picking up the hook to try it again.

"Can you get that hook under the pedicle?" Richard said.

"Yes, there, it's in. I'm going to hold it in place this time while you attach the rod."

Richard attached the rod.

"Great," I said, "now for that bottom hook." But as I notched the hook onto the bottom vertebra, the spine shifted and the whole thing fell apart again. This went on for hours. From the other seven tables explosive utterances echoed out: *Damn! I thought I had it this time.*

"Put your finger here," I said at one point, "while I put this nut over the end of the rod . . . Hold the rod stable while I tighten the ratchet." And so it went. We were lucky some days to get one rod in before lunch and the other by the end of the afternoon.

Throughout it all, Sandy Larson circulated throughout the room, a gentle, soft-spoken man with a professorial air. A phenomenal teacher, he adjusted a rod here, a hook there, always teaching by doing, always encouraging us. I'm sure that when I came away from the course, it was not only instrumentation I had learned—as valuable and career changing as that was; I'd also witnessed the humane hallmarks of a great educator.

Upon my return from Milwaukee, I officially began to do instrumented thoracolumbar surgeries. That—along with my increasing political involvement—was tantamount to pinning a big red target to my chest. Dr. Zimtbaum's request for that case review was only the first broadside.

About a month after the review with the orthopods, Robert welcomed me into his office. Late afternoon light made a pale, cool rectangle of the frosted glass window to the east. A Mozart sonata filtered softly from a speaker in one corner, and on his desk a TV monitor showed a view of an empty OR 1. He switched the monitor off, held up a letter, and motioned me to sit down.

Like me he was dressed in the standard blue scrubs. The tan he'd gotten snorkeling in the Caribbean a month earlier was just beginning to fade. Ordinarily we might have bantered at this time of day. Catch up on some family news or sports. Then I'd head home, feeling relaxed and good about things. I knew this was not going to be that kind of visit.

"Thought you'd want to see this right away, Volk," he said.

"What is it?"

"Let's just say if you have any instrumented spine surgeries scheduled, you'll need to make some adjustments before you do them."

"Who's it from?" I asked. "No, let me guess. John Bull."

"You got it. Copy to Dr. Zimtbaum."

"Another case review," I said.

"No, not this time, but just as annoying."

I felt my stomach kick. "Well let's have it."

"Orthopedics have convinced the executive committee to lay down some restrictions—they're only stipulating one now, but more are sure to come—on how and when you can operate using instrumentation. This includes Richard Douglas by the way, and the other fellows."

"Just what kind of restrictions are we talking about?"

"Well, they know they can't stop your doing thoracolumbar, but they want one of their own to supervise the surgeries you do."

"You've got to be kidding," I said.

"I'm afraid not. I've made it quite clear that I don't approve of any extension of the review process and that I consider the reports biased and inaccurate and their criticisms arbitrary. So, now they're taking a different tack."

"I don't see how the hospital can agree to this. Sandy Larson's course on instrumentation has accredited neurosurgeons all over the country, at reputable institutions. And our results have been good. We've had none of the most severe complications. No permanent neurological deficit. No deaths. Not even a friggin' poke with a sublaminar wire."

"I know that and you know that. They know it too. It's political. I'm not going to back down. But for now we're just going to have to beat them at their own game."

I shook my head. "When does it start?" I said.

"Effective immediately."

That first edict was soon put to the test. Richard did the case with me, this time a tumor of the thoracic spine. As we stood scrubbing our hands, we went over the main features of the operation.

"OK Chiefy, it's going to be a marathon today. We'll come in anterior to get the tumor, then turn her over to repair the spine. We'll have to be extra careful with the sublaminar wires, and where we place them will depend on how much damage has been done to the columns."

"We'll be lucky if the tumor has damaged only two of the columns," Richard said.

I gave a shake of my head. "I wouldn't count on it."

"Right chief," Richard said. Then, "Any idea who the orthopod is today?"

"No, I didn't check," I said. "Whoever it is, we'll just do what we have to do."

We pushed through to the OR and allowed the scrub nurse to gown us. The anesthesiologist—an impressively muscled black body-builder named Ed Washington—had already "passed gas," and the patient— an ICU nurse from Maricopa County Hospital whose breast cancer had metastasized—was now asleep, her chest exposed and ready for the incision. The scrub nurse waited to one side, and the circulating nurse stood near the door.

We waited. Soft Rock music mixed with the bleeping of the EEG and the rhythmic whoosh of the oxygen. Richard and I looked at each other, cocking our eyebrows. *Where was the orthopod?*

"Can someone call staffing?" I said. "See who's scheduled and ask if he's left yet."

The circulating nurse darted out and came back a few minutes later. "It's Dr. X," she said. "He's on his way."

Ten minutes later Dr. X still hadn't arrived. "Call again," I said.

The nurse left and returned. "They said he left fifteen minutes ago."

"What do we do Chief?" Richard said.

"I'm sure as hell not going to wake the patient up and put her through the prep again tomorrow," I said. "If he's not here in five minutes, we'll go ahead."

The orthopod never did show up and we went on without him. He missed a good opportunity too. It was a very long and tedious surgery. The tumor had eaten away at the anterior and middle columns, destabilizing the spine. Once we removed what was accessible of the malignant growth, we placed a bone graft from the patient's hip into the empty cavity where the tumor had been. Then we turned the patient over and went in from the back to remove the remaining tumor and do the fusion. We used Harrington rods, one on either side along the facets, securing them with sublaminar wiring. The wires were stiff in those days, and passing them so near to the spinal cord took a great deal of control and skill. One ding and you could cause serious damage. Finally we threaded the ends of each wire through a needle holder to twist them tight. The instrumentation alone added about three hours to the operation. And though the removal of the tumor was successful, as well as the instrumentation and fusion, an infection did set in. I had to take the patient back into surgery, clean all the pus out, and put what we call a "feed-me-drain-me" tube in. That gave me some sleepless nights, knowing the orthopods were watching me like a hawk. But the patient did fine after that.

"Harrumph," I snorted to Richard while we scrubbed out. "The orthopods have gained the field but now they won't even play."

The battle slogged on without resolution. In early fall, a second edict followed the first: I could do the procedures, but I would be required to have an orthopedic surgeon as co-surgeon. Then another: the case would have to be reviewed by a committee of orthopedic surgeons before I could perform the procedure to ensure that I was using the appropriate instrumentation. The environment grew increasingly hostile, and the mudslinging continued. I tried to take it all in good stride, but the doubts of my colleagues gave me some angst. The poisonous climate was wearing some of them down.

One Friday about a month after my unsupervised surgery, I came out of Grand Rounds to find a gray-haired, senior member of the neurosurgical group at Good Sam waiting for me. He was obviously disgruntled. "I just want to know why you are doing this, Sonntag?" he said.

"For years we did the decompression and the orthopods did the fusion. Why not let it stay that way? It's worked fine. We really have no business doing instrumentation."

"I'm sorry to hear you feel that way," I said. "As you know, organized neurosurgery thinks that we should be teaching our residents thoracic lumbar instrumentation. The only way we can do that is for us to do it ourselves."

He smacked his lips as if he had a bad taste in his mouth. "Well, you're just creating divisions," he said. "Bad feelings all around."

Another time one of our own associates caught up with me coming out of the cafeteria where I'd gone to grab a cheese sandwich. An amiable, dark-haired Irish guy, he was obviously pained by the conflict. "Your doing thoracic-lumbar instrumentation is causing all kinds of trouble," he said. "And it's dividing the orthopedic and neurosurgical communities."

"Gee, I'm sorry for the trouble," I said, and I meant it. "But someone has to teach the residents how to do these procedures. That's why I went to Milwaukee to take Sandy Larson's course. You know, I asked the orthopedic surgeons to teach me the procedures themselves. But they wouldn't do it."

He nodded his head in glum agreement.

"And," I added, "we are a teaching institution. We have to move forward."

Despite the few objections, I felt the rest of the local neurosurgeons were just keeping an eye on the ring to see if I got knocked out or stood up again. *If that's what Sonntag wants to do,* they seemed to be thinking, *let's see what happens.* In the end, I held my ground. There was too much at stake. A neurosurgeon simply had to be able to treat the whole spine—that was the definition of neurosurgery, taking care not only of the brain but the spine, the peripheral nerves, and all the bony structures that encase those crucial neural pathways—and if we were going to do that, it was essential that we be allowed to teach residents how to do these procedures as well. It wasn't only my future I was fighting for; it was theirs.

On a cold evening in December 1990, Robert Spetzler arrived at my door holding a large sealed Manila envelope. A great fir tree twinkled with white lights just beyond the entryway, but the house was quiet. Lynne had given birth to our third child—Stephen—just days before and was occupied with the three children. "The results are in," Robert said,

In 1991, when this photo of our family was taken, the controversy with the orthopedic surgeons was reaching a breaking point. Lynne and the children kept me sane.

waving the envelope. I felt my lips press together in a grim line. I motioned him in.

We didn't speak until we reached my office upstairs. Only when I'd closed the door did I face him. "Well?" I said.

"I thought we should find out together."

Robert gave me the envelope. I ripped it open and pulled out a thick sheaf of papers. I looked at Robert. I knew whatever the contents were he'd back me up. He'd done so through the whole nasty business. But I keenly hoped I wouldn't let him down now.

The envelope held another set of evaluations. They would determine either the end of my venture into the spine or leave to carry on the struggle. With no way through the impasse between the orthopedic and neurosurgical communities, the hospital executive committee had finally suggested that independent reviewers evaluate my first twelve thoracolumbar cases. This time, however, the executive committee included a representative from Barrow. That representative was Robert Spetzler.

"That's fine," Robert had said. "Take Sonntag's twelve cases for evaluation, but then, why don't you submit for review twelve orthopedic cases as well, whichever ones you want. Then we'll run a blind review of all of them and see how they measure up."

The committee had agreed. Now, after months of brooding, I held the outcome in my hand. I flipped through the first pages, all reviews of my cases. A negative comment here and there, but overall positive. Then I scanned the reviews of the orthopods' cases, handing them to

Robert when I'd finished. Slowly the news sank in. When I looked up again, Robert was grinning at me. "You did it, Volker," he said. "You beat them on their home court."

And so I had. The orthopedic cases had received more criticism than mine. I knew the war wasn't over, but I had won my Gettysburg.

CHAPTER 14

THE PEDICLE SCREW EXPOSÉ

The first thing we do, let's kill all the lawyers.
WILLIAM SHAKESPEARE

IT WAS A THURSDAY evening in 1993 and I had just tuned in to ABC TV's investigative program *20/20* when the phone rang. It was Steve Garvin, an orthopod friend of mine, then chair of the Department of Orthopedics at UCSB. We kept in touch but I immediately wondered what the heck he was calling me about at home.

"You got the TV on?" Steve said. "Tune in to ABC's *20/20.* You've got to see this."

"Yeah, hold on," I said. "I just turned it on." I turned the sound up as hosts Lou Downs and Barbara Walters appeared on the screen. The funereal voice-over jolted me into riveted attention: "Tonight the medical device they say ruined their lives . . . back screws inserted into their spines."

I heard Steve sputtering on the phone. "Do you hear that? Can you believe it?"

"I'll call you back," I told him.

Then I watched as the hosts revealed that they had uncovered "shocking facts" about a widely used medical device the FDA ostensibly considered experimental. Cannily interspersing the testimony of experts with sound bytes of patients who claimed their lives had been ruined by these "back screws," Downs and Walters peppered the program with inflammatory references to patients being used as "guinea pigs" by "quack doctors" with financial ties to the primary companies who manufactured the device.

I called Steve back after the program. "Well, the lawyers must be dancing now," I said.

"You bet," Steve replied. "I expect we'll see a new crop of suits."

"Yeah, you're right," I said. "Those vultures will never stop now."

"What's the climate like at the associations?"

"Mixed. This thing's going to stir things up even more. It's the main item on every agenda."

"Sonuvabitch," Steve said.

"Well, let's not get our trousers in a twist just yet," I said. "The debate over these new devices has been going on for a while now. We'll have to wait and see what the repercussions are from this 20/20 thing."

Many of us had reason to worry. Various medical device manufacturers—Medtronic was a big one— and the main professional spine associations had been fighting a legal battle for two or three years over the device. The controversy was well known to physicians working on the spine, with some advising caution and others wanting to forge ahead with the new instrumentation. Lots of people were getting jumpy. I was feeling a little paranoid myself, wondering what was being said behind my back: *Watch out for Sonntag. He's going to have a lot of lawsuits from putting in these instruments.*

Along the way, the controversy had become fodder for the media. I myself got a call from a *20/20* reporter months before the broadcast. They were looking into the pedicle screw issue, the reporter said. I didn't know then what angle they were working, but I soon sensed the reporter fishing for a negative take on the orthopods. Did I think the orthopedic surgeons should be performing surgeries with the pedicle screw? Why should orthopods operate near the nerves?

I certainly wasn't about to badmouth the orthopedic surgeons, and my responses were so neutral that when the program eventually aired, it featured no sound bytes from me. You've got to give *20/20* credit though. Their smear campaign took things to a dramatic new level.

The aftermath of the *20/20* pedicle screw broadcast aptly illustrates the American phenomenon known as mass litigation. The broadcast fueled an up-spike in the ongoing legal action. It also spurred a flurry of plaintiff newspaper ads that sprouted like mushrooms after the rain all across the country. (The one that ran in Arizona brazenly read: DO YOU HAVE SCREWS IN YOUR BACK? CALL THIS NUMBER, not mentioning anything about pain or complications.) Hundreds of new lawsuits were filed. The plaintiffs' legal representatives alleged that these devices were inherently defective, that their experimental standing had not been divulged to patients, and that because they were in an experimental phase, any use of them constituted malpractice.

Unfortunately, while the 20/20 story certainly made for good entertainment, as such it either ignored or distorted information about the history and development of what they insisted on calling the "bone screw" for use in spine surgery.

The pedicle screw had been developed in the 1960s by Canadian, American and French surgeons to stabilize the unstable spine. Harrington had been an early user, finding that pedicle screws, in concert with the rods, attained a much better fixation than the original system using hooks. Anchored in the bone rather than lying on top of it, the pedicle screw soon became standard. With advances in instrumentation, to add the pedicle screw to the spine surgeon's armentarium—the array of tools and procedures from which a physician chooses the most appropriate for the case at hand—was a natural progression. While the FDA had sent letters to six manufacturers earlier in 1993 stating that they could not "advertise for promotion" the use of bone screws as pedicle screws, they did not ask the companies to stop manufacturing the device nor did they prohibit their "off label" use in spine surgery.

"Off label" use meant that the patient needed to be informed that a medication or device (in this case the pedicle screw) had not yet been approved by the FDA but that the surgeon believed that in a particular patient's case, it was the right instrument to use. At Barrow, we took the requirement that patients be informed seriously. Any patient for whom the pedicle screw was recommended received a three-page letter explaining in detail why the surgeon wanted to use the device even though its status was still Class 3, or "off label."

It was true that the benefit of pedicle screws was not yet well documented, and to correct this deficiency, the FDA appealed to a group of spine-related specialist societies for assistance in developing a study, the results of which could possibly justify the reclassification of the device to Class 2, that is, FDA approved and not off label. The study, started right around the time 20/20 aired the damning program and completed in 1994, was organized by a scientific committee composed of spinal surgeons, and solicited input from the five principle societies concerned with the spine: the American Association of Neurosurgeons (AANS); the American Academy of Orthopedic Surgery; the Scoliosis Research Society; the Congress of Neurosurgical Surgeons (CNS); and the North American Spine Society (NASS). The spinal implant manufacturers group, consisting of fourteen companies, underwrote the cost of the study. Organized neurosurgery supported the study and encouraged neurosurgeons to participate. Sensitive to any accusations of conflict of interest, and to encourage fair reporting of both good and bad outcomes, the FDA assured participating neurosurgeons that the reports would be confidential.

By the time the study got underway, however, the litigation engine was in full throttle, and once the federal court had designated the cases as "multi-district litigation," and bestowed on the plaintiffs' lead counsel the role of Plaintiffs' Legal Council, or PLC, the attorneys began the process of discovery. As part of this process, they attempted to obtain the names of the surgeons and patients who had participated in the cohort study.

Confidentiality is a key component of most research protocols, and patients and physicians participating in the study had been given assurances of confidentiality. The AANS, the Orthopedic Academy, and NASS together filed a motion to intervene to protect the confidentiality of this information, a motion that was fortunately granted. The ball was back in the plaintiffs' court. Within weeks, the AANS and the other organizations were named as defendants, with the PLC alleging a series of nefarious acts on their part: the associations had acted as "promotional centers" for the pedicle screws; they had engaged in "reckless and outrageous" promotion of hazardous spinal devices; and had "conspired with manufacturers for the illegal sale of dangerous medical devices."

Between the manufacturers and the big associations, the lawyers had found their deep pockets.

The two men were typical lawyers. Dressed in uniform suits and ties, the only difference I remember was their height; one had oddly truncated limbs and stood a foot short of his partner. At their side was a pillow-shaped woman wearing a plain skirt and jacket, low shoes, and glasses—a local court reporter I soon found out. The two lawyers had flown out to Phoenix on behalf of the PLC, who had subpoenaed me as a representative of AANS, the American Association of Neurological Surgeons. Backing me up was an association lawyer. The five of us met in the lobby of a Hilton hotel near the airport. I felt the urge to wipe my hand on my suit coat after the requisite handshakes.

The subpoena itself had been short and to the point, commanding me "to produce for inspection and/or copying any and all documents, records, or other tangible things in my possession or under my control" relating to what they hoped to show was collusion between the spinal implant manufacturers and the spine associations—and to appear at the designated time and place to give a deposition.

I'll admit to some alarm upon receiving the subpoena. I knew the lawyers would come with information they'd dredged from the manufacturers' and association documents through further subpoenas. Still, when the day came, I felt, if not calm, resolute. Anyone who knows me knows that honesty is one of my mainstay values. I would answer their questions as truthfully as I could, without volunteering anything they could twist.

We rode up in the elevator together, silently looking at the ceiling, and entered a small conference room with a table for six. The AANS lawyer and I took seats with our backs to the window. The distant roar of an airplane taking off broke through the shuffle of the lawyers' preparations. The court reporter swore me in. I leaned back and crossed my arms. The tall lawyer opened.

"As you know Dr. Sonntag, we are here today to conduct discovery regarding legal action taken by the PLC against the medical device manufacturers and spine associations. Today we will focus on your role in the American Association of Neurological Surgeons. We will be asking you questions about the facts you have in your actual possession regarding the issues in the law suits." This was followed by a long general statement of the rules: a reminder that I was under oath; directives about answering with a vocal "yes' or "no" instead of a nod; and an advisory against guessing, wholly neutralized by their encouragement to give my best "estimate" regarding dates, times, etc.

I indicated my understanding and the grilling began. They dispensed with my background in short order—no ex-wife holding a grudge, no egregious failures or dismissals—and moved to the case at hand. The first topic was the AANS national meetings.

"Dr. Sonntag, you are very active in AANS. What has been your role in organizing the national meetings?"

"What role I have centers around the cervical instrumentation course I teach at the meetings," I said. "Other than that, I have nothing to do with the arrangements."

"I see. At these meetings, is it customary for representatives from the manufacturers of cervical plates and screws to attend?"

"Yes."

"Who are these manufacturers?"

"Medtronic, Sofamor Danek" I said, adding the names of other companies.

"How is it decided which manufacturers may attend these meetings?"

"All the manufacturers that make spinal implant devices, in particular cervical instrumentation, routinely attend these meetings."

"How are they notified?"

"When I have the course ready, I have my secretary call the companies that make the device I might be focusing on—cervical plates for example—and make sure they have the course date and time, and bring their instruments so we can demonstrate them."

"And you call all of the manufacturers."

"My secretary calls them. We invite all of them equally."

"And they all have an opportunity to demonstrate their devices."

"Yes, that's right. We set up ten tables or however many we need for those manufacturer reps attending, and at each table a spine surgeon demonstrates the device."

"Doctor you've run the cervical instrumentation course at the national meetings for many years. Is that correct?"

"Yes, that's correct," I said.

"In your instruction Dr. Sonntag, have you favored one type of instrumentation over another?"

"No, I have not."

"But isn't your role to advise on the best devices to use for particular surgeries?"

"My role is to familiarize the attendees with the instrumentation available for a particular procedure, and to discuss the pros and cons of each."

"At the national meetings, did you direct attendees of your classes to the tables of particular manufacturers?"

"No, I did not."

"Dr. Sonntag, have you in any way endorsed one or more of the manufacturers, at the meetings or in association publications?"

"No."

A pause ensued, during which the court reporter tapped away at her machine like a woodpecker. The two lawyers conferred and rifled through some papers. The short one straightened his tie and cleared his throat.

"Dr. Sonntag, can you describe to us your relationship with Ron Pickard?"

"Ron is the president of Sofamor Danek. Naturally I know him from the national meetings."

"Would you say you are friends with Ron Pickard?"

"In a professional sense, yes."

"Have you socialized with Mr. Pickard?"

"No, not outside of work-related functions."

"I don't think I understand you Doctor," he said. "You have social-ized then with Mr. Pickard?"

"No, I have not socialized with Ron Pickard in the sense that you mean. I have not met with him other than at the national meetings."

The questions went on in this vein until lunchtime. We broke and went our separate ways for an hour.

"You're doing great," the association lawyer said when the elevator doors had closed.

"I'm telling the truth," I said. "They think by asking the same ques-tion over and over I'm going to change my mind. Or my memory will change. Or maybe they'll wear me down and get me to frame my answer in a way they like."

"That's a typical tactic. It's all part of the process of discovery."

I looked at him and thought, *Right, discovery. More like entrapment of the unwary.*

"It's clear what they're fishing for," I said, "That I was somehow try-ing to influence other neurosurgeons to use one particular product over another. To anybody who knows me, the idea is ludicrous. It's not my style to hawk a product."

"We know that. Your testimony today is not going to give them any-thing they can hang a hat on."

After lunch the grilling continued. Now they tried a new tack.

"Dr. Sonntag," the tall one said, pulling out a document. "Your name appears on several lists of physicians invited to go on all-expenses-paid trips sponsored by various manufacturers. Were you aware that your name was on these lists?"

"I was aware that trips were planned because I was invited, but I don't know about any lists."

"But you knew of these trips and you were invited to them."

"Yes."

"And these trips were categorized as 'think tanks'?"

"Yes, that is correct."

"To your knowledge, did many neurosurgeons go on these trips?"

"You would have to ask each physician in question about that."

The lawyer picked up another paper, smirked at his partner, and waved it in my direction. "Doctor, these records show that you were invited to one of these so-called think tanks hosted by Sofamor Danek in Alaska. Is that correct?"

"Yes, that's right."

"And here's an invitation to one in Montana sponsored by Medtronic. Do you agree that you were invited on this trip?"

"Yes, I was invited on a trip to Montana."

"And both of these trips were all-expenses-paid?"

"Yes, that was my understanding."

"And yet you claim to be impartial when it comes to recommending the use of medical devices to your colleagues in AANS and other spine associations."

"Yes, that's correct."

"How can you say you're impartial, Doctor, if you've accepted invitations from the manufacturers to go on these trips? Surely you must have been influenced by the company sponsoring a particular trip."

I waited a moment to answer. The room felt thick with predation. Then I raised my eyebrows and looked each one in the eye. "Who said I went on the trips?"

Another shake of the documents. "But we have here copies of the invitations."

"Just because I got an invitation doesn't mean I went. I didn't go on a single trip."

The lawyer fell back in his chair. On a purely sporting level I could almost sympathize with him. He thought he'd caught a big one. And he certainly would have had a decent catch for his efforts that day if I had been so clueless or unethical as to accept those offers. Because the two lawyers did get that right. There had been plenty of invitations to go on these thinly disguised jaunts. Of course I'd known what the medical reps were trying to do. It was so obvious. And I hadn't wanted any part of it. It wasn't that I thought I might be sued down the line. I had no idea about that. I simply said at the time, "Come on guys, give me a break. Attend a think tank meeting and fish all day?"

In the end, the PLC got nothing out of me that they could use. I bolted out of there around sunset, drove home, changed and picked up my oldest son Christopher. After two hours of coaching his soccer team, of running and sweating with the boys, I felt clean again.

It did not look good. With the PLC mobilizing for a big offensive, the AANS, NASS and the Orthopedic Academy carried on the fight for re-classification of the pedicle screw, now with the support of the manu-facturers, in particular, Ron Pickard, the president of Sofamor Danek. Pickard was a hard-working, self-made Midwesterner who had started off as a farm boy and risen to the head of one of the biggest makers of spinal and cranial surgical technologies. (Sofamor Danek later became part of Medtronic.) He was a major force, if not the major force, in the fight against the PLC.

As for the individual spine surgeons, none of us were personally sued, but once we became aware of the forces conspiring against us, we felt more like brothers in arms than combatants in a turf war. We realized how 20/20 and the PLC had been playing both sides. My own epiphany came some months after my phone call from the 20/20 report-er. I was on the line with the orthopod friend of mine who had called me the night of the 20/20 airing—Steve Garvin—when I remembered that exchange with the reporter.

"Hey, Steve" I said. "Did you happen to get a call from 20/20 some months back?"

"Yeah," he said.

"What did they want?"

"They wanted me to badmouth the neurosurgeons. They wanted to know how neurosurgeons could do all that bone work."

"No kidding," I said. Then I relayed my own experience with 20/20.

"Sons-o-bitches," Steve said. "That is a sad, sad situation."

The legal tussle went on. NASS, the AANS, and the Orthopedic Academy held their ground, and then slowly advanced through lob-bying, legal counsel, subpoenas, and meetings. The leadership of the AANS would change four times before it was over, its successive pres-idents—Doctors Stewart Dunkser, Sid Tolchin, Russell Travis, and es-pecially Ed Laws—taking up the fight in turn. And while one or two manufacturers folded to the pressure—one named Acromed agreed to pay the PLC $100 million as long as the company was not named in a suit—the others held tight. Together the manufacturers and associations named in the suits sank thousands of dollars into their defense, but for a long time, the plaintiffs only gathered more steam. Some lawsuits were tried and plaintiffs were winning the first round. In Texas, a jury returned a verdict against Sofamor Danek for the amount of $414,000.

Finally, four years after the study began, the results came in. That was the turning point. The data conclusively demonstrated that the pedicle screw was as safe and effective for the surgical treatment of spinal fractures as other internal fixation devices. In fact, the results from 314 surgeons treating 3,498 patients showed a 90% rate of successful fusion in patients whose treatment for degenerative abnormalities of, primarily, the lumbar spine included pedicle screws, compared to 70% of control patients who achieved fusion but did not receive pedicle screws. Those results were good enough for the FDA Advisory Panel on Orthopedic and Rehabilitative Devices, who recommended that pedicle screws be reclassified from Class 3 to Class 2. The final step was for the FDA to translate that recommendation into policy, which they did on July 27, 1998, reclassifying the pedicle screw to Class 2.

After this reclassification and a growing number of verdicts against the PLC, most of the plaintiffs' suits were dropped. The PLC made one last desperate charge and filed suit against the FDA, seeking to prohibit the reclassification of the pedicle screw, but there was no wind in that sail and that suit, too, was eventually dropped. Fortunately for the lawyers, things were heating up with an anti-obesity drug called Fen-Phen around that time, and many of them darted off to get a piece of that action—at an estimated liability of $14 billion, Fen-Phen turned out to be a far bigger fish than the pedicle screw.

The pedicle screw issue did not completely repair the rift between the orthopedic and neurosurgical communities, but it did stop it from widening. In face of the PLC's attack on our professional ability, integrity, and moral authority, the previous struggle over instrumentation lost its urgency. After all, both groups had exactly the same goal at heart: the best possible care of patients with spinal disorders. So, we all did the Kumbaya thing, shook hands and buried the hatchet. In the mid-1990s, even before the final verdicts of the courts and the FDA, the Council of Spinal Societies (COSS) was formed, thus bringing together in one organization all the major organizations involved in spine care.

For me, the victory was one more vindication of the path I had chosen. I was thrilled that now not only neurosurgery but also the other specialties acknowledged that neurosurgeons were capable of treating the complete spine and performing complicated instrumentation procedures on every area from the occiput to the sacrum, both anteriorly and posteriorly. And though it took several more years, eventually the American Board of Neurological Surgery added treatment by fusion and instrumentation to the definition of what a neurological surgeon does.

CHAPTER 15

PHYSICIAN TO A QUEEN

It is health that is real wealth and not pieces of gold and silver.
MAHATMA GANDHI

LYNNE TOOK THE call for me. It was October 1994, the year of the O. J. Simpson trial, and I was attending a conference in Cartagena, Colombia.

"Someone called from Paris," she said when I returned to the room. "They want you to fly out there to consult on a case."

"I can't do that," I said. "I've got a meeting here, and I have to give a talk tomorrow."

I left to attend another session, but when I came back from the meeting, they had called again.

"They really want to talk to you," Lynne said. "Something about consulting on the wife of the king of Saudi Arabia."

I called the number back. Sure enough, it was the Saudi royal family. They wanted me to fly to Paris immediately to examine the "queen." (The king had two wives, and I am not sure how they distinguished between them, but that is how her associates referred to her in our discussions, and that is the title the press eventually used, too.)

"I can't come out there," I told them. "The meetings aren't over here until Friday."

"Well, can you come on Friday then?" they asked.

"I already have my ticket back to Phoenix," I said.

"We'll take care of everything," they answered.

So, on Friday Lynne headed back to Phoenix, and I got on a plane to Paris via Bogotá.

A limousine picked me up at Charles de Gaulle Airport and ferried me into Paris, where the "House of Saud" had arranged my stay at a five-star hotel. After the fourteen-hour flight and 45-minute ride, all I wanted was to shower and rest. I had just enough time, though, to splash my face, change, and glimpse the Eiffel Tower through the sheer drapes before the chauffeured car reappeared to deliver me to the queen. I had no idea what to expect, and I wasn't even sure what *arrondissiment* we were in when we passed through a large gate set into high walls. The car crunched up a long gravel drive, wound around, and stopped at what I gathered to be the main house of a palatial compound. Two men, one of them a short, swarthy, heavy-set man with greased-back hair, came forward to greet me. "Welcome, welcome Dr. Sonntag," he said, shaking my hand and introducing himself as Dr. Khouri, the queen's main physician, before ushering me inside.

We passed between two stone-faced guards posted at the entrance and entered a reception area the size of a hotel lobby. My feet sank into a mammoth Persian carpet covering a highly buffed marble floor. Baccarat crystal chandeliers sparkled overhead, their light burnishing gold-leaf moldings and flowered ceiling appliques. Ornate French doors led off to various rooms. My greeters—the second man was another of the queen's physicians—steered me up a sweeping staircase and into a spacious sitting room. There I met the man I came to know as Prince Number One.

He appeared to be similar in age to me with dark, thick hair and deep-set eyes. Possessing a fine olive complexion and well-defined lips, his full, almost plump, face was relieved of any boyishness by his polite but commanding air. He spoke little to me, and had the short, stocky physician—Dr. Khouri—interpret when he did.

We wasted no time getting down to business. Dr. Khouri filled me in on the queen's history of serious medical conditions—the most recent a neck injury sustained in a car accident in Paris that had left her with weakness in her arms and legs—and handed me some MRI scans of the queen's cervical spine. I walked over to a window and held the films up to the afternoon light. The severity of her condition was immediately evident by the curvature of her neck; instead of curving slightly back, the spine was frozen in the reverse direction—a condition referred to as a "swan neck." In addition, bone spurs had developed on the vertebrae, narrowing the spine and compressing the spinal cord.

The prince lit a cigarette, took a long drag on it, and spoke in Arabic to his interpreter.

"His highness would like to know your diagnosis," Dr. Khouri said.

"Well, this reverse curvature here indicates cervical kyphosis. See how rigid the spine is? Also severe stenosis of the cervical spine—this thickening of the bone. See these bone spurs here and here. That's causing a narrowing of the spinal canal, which can damage the nerves. You can see the bone pinching the spinal cord here."

The prince conferred with the doctor and nodded.

"We will bring her highness now," the doctor said.

A moment later a veiled and robed woman wheeled a small swaddled figure into the room. Two or three other women followed in what looked, to my Western eyes, like nuns' habits. I was introduced. The queen nodded, and through the veil I heard a gentle voice speaking in Arabic.

"The queen welcomes you and hopes that you had pleasant travels," the interpreter said.

I thanked the queen, wondering how I was to examine this woman; she was covered head to foot in a dark, robe-like garment with intricate stitching and a long veil pulled low over her forehead. An additional piece of handkerchief-sized fabric fell from just under her eyes to the middle of her chest.

I turned to the prince and his interpreter. "May I proceed as usual?" I said. The prince nodded his agreement. With Dr. Khouri interpreting, I ran through the usual steps, telling her to wiggle her toes and fingers, and asking her where and how severe her pain was. I instructed her to push against my arms with hers to test her strength. One of the doctors handed me a small hammer. I moved the fabric of her robe and tapped her knees to test her reflexes. I took a pin and pricked her arms and legs. Then I pulled aside her veil and examined her neck. None of her responses surprised me.

The attendants wheeled the queen out. Charged (as far as I could tell) with preventing the queen from lifting a finger, the silent, veiled women had stayed constantly by her side. Once she had left, the prince spoke through his interpreter. How would I treat the problem, he wanted to know. I returned to the window with the MRI scan and explained to him and the doctors what I would do: a four-level discectomy followed by fusion using a plate to stabilize the spine.

"Thank you Doctor Sonntag," Dr. Khouri said. "That will be all for now. Please, go back to your hotel to rest. We will send a car for you again this evening."

I returned to the hotel and collapsed on the bed. I thought of Lynne and wished she were with me. She would have enjoyed the room: fancy brocades and period furniture, a lot of gilt. It was like a chamber Marie Antoinette would have inhabited. I felt too wired and anxious to sleep. I took a short walk and tried to throw off the heavy blanket of jet lag blanketing my head.

The car returned for me about six o'clock. This time I was led through a different set of French doors into another luxurious sitting room. Four men already sat there; they turned their heads like cogs on the same wheel when I entered. I recognized all but one—a doctor from Belgium or Holland. The others included an orthopedic surgeon from Paris, a neurosurgeon/orthopedic surgeon named Andre Wagner from Hamburg, and an American neurosurgeon named Don Long. I took my place among them, feeling like a comrade at arms.

"Hello Volker," Don said. I see you made it to the finals."

"Looks that way," I said. "Pretty tense process, eh?"

"A circus," one of the others said.

"You can say that," another said. "Maybe a big headache in the end."

"I hear they started with a list of a hundred or so, Don said. "They whittled it down to ten to actually fly out to examine her."

"Where are the others?"

"Gone now. Alan Crockard was here earlier in the week. He didn't make the last cut."

"We're it then?" I said. "The Final Five."

The others shrugged and sighed. *I hope to God they don't pick me*, they seemed to be thinking.

We lapsed into intermittent observations. The door opened and a servant entered with a tray bearing coffee in china cups. We sipped and shifted in our seats. A half hour passed. I wondered how much longer it was going to take. I felt hot and remembered sweating on the plane. I pulled a handkerchief out to wipe my face.

"You okay?" Don said.

"I don't know," I said. "I think I might be coming down with the flu or something."

My breath came shallow. *Damn*, I thought. *Am I having a panic attack?* The next minute a sharp pain stabbed me in the gut. I felt a sudden urgency and squeezed my sphincter tight. An image of shrimp and seafood cocktails flashed in my mind. *Oh no*, I thought. *Not now.* Balefully I recalled the stopover Lynne and I had made with the Spetzlers in

Colombia two days before. The exotic island, the boat provided by our host, the scuba diving . . . the pounds of shrimp and seafood glistening in the sunshine. My stomach lurched. The sweat came on in force.

"Excuse me," I said, getting up. I hurried out into the reception area and its dozen doors. Thank God—someone to ask where the bathroom was. I was directed to a kind of vanity area—another acre of marble floors—with yet more doors. Light reflected off myriad gold fixtures, but I wasn't in a state to appreciate them. *Where the hell was the friggin' toilet?* I opened door after door: linens, supplies, toiletries, a massage room. Finally a bidet. *To hell with it,* I thought. *That'll have to do.* I charged in. I sat down. From my new vantage point I noticed another door ajar. There, partly hidden, gleaming gold, was the unmistakable shape of a toilet base. I pulled my pants up, hobbled over, and exploded on the golden throne.

After composing myself, I returned to the sitting room. Before long, a man came in and addressed the group. "Thank you Doctors. We very much appreciate your time. You may go now. The cars are waiting for you." I think I sighed with relief. But as I moved toward the elevator with the others, I felt a tug on my sleeve. "If you would, Dr. Sonntag, please stay," the tugger said.

As the others filed out, I was led into a room dominated by an enormous table with, again, the syrupy luster of gold. Around it, like dignitaries at a banquet, sat a dozen individuals of both sexes in Arabian attire—I later found they were members of the Saud family and their associates. I took the empty seat indicated, across from Prince Number One and Dr. Khouri. With a nod from the prince, the doctor spoke.

"The family have selected you to do the surgery," he said. "How soon can you do it?"

"Well, I would need to go back to Barrow and make the arrangements," I said.

"How about Wednesday?" he said.

"Wednesday? Today is Saturday."

They were unfazed. "The embassies in LA and New York are already working on it," he said. "They will handle all the formalities."

By now, my nervousness had condensed into another profuse sweat. A series of questions about the surgery followed: *How would I operate? What were the dangers as to the queen's flying out to Phoenix?* I explained that the examination had revealed a tight spinal cord and that any sudden bumps during the flight could cause the queen to stop breathing.

That alarmed them, but they found a ready solution. I would simply have to arrange for a neurosurgeon and an anesthesiologist from Barrow to accompany her on the flight from Paris to Phoenix.

Once they had exhausted their questions, only one detail remained. The doctor stood, indicated for me to follow him, and led me to the door.

"As a final step, Dr. Sonntag, please tell us your fee."

"Well, I don't know. It's so unusual," I said. "Five hundred . . ."

"Five hundred . . . thousand?" he said, without blinking.

No," I said, "five hundred dollars." I figured they had already flown me to Paris first class, and I didn't want to be greedy.

I walked out. Near the entrance stood a table on which sat a box of envelopes. A man was seated there. The doctor whispered something to the seated man, who then reached over, picked up one of the envelopes and handed it to me. I put it in my pocket.

The limousine picked me up the next morning and took me back to Charles de Gaulle. I boarded my flight to New York with a connection to Phoenix. Somewhere over the Atlantic, I remembered the envelope. I opened it up. There was ten thousand dollars in cash inside in crisp hundred-dollar bills.

I raised the plastic shutter and peered out the window. Far below, clouds skiffed over the expanse of deep blue. I never tired of such a view. I wondered if the troop transporter my family had shipped over on had traveled those same currents. Despite the drone of the airplane, I felt calmer. My stomach had settled down. I was relieved I'd been able to pull through the jetlag and cramps and focus on the queen's case. That kind of physical control is one of the unheralded skills a doctor picks up along the way. My thoughts turned to an earlier time when my control was tested in the middle of an important licensing examination.

It had been late in my residency, when I'd had to deal with the problem of not having passed the first part of the USMLE by taking an all-day exam called the ECFMG (Educational Commission for Foreign Medical Graduates). This exam was designed for foreign doctors who wanted to be licensed in the US, but American residents who had not passed all three parts of the USMLE could also take it.

I'd studied like hell for that exam. It covered all areas of medicine, so even though I was in the neurosurgical training program at Tufts, and was doing a six-month neurology rotation at Boston University Neurology Department, I had to review all the other areas of medicine to pass it. I needed it to be licensed in the state, but I also wanted to redeem myself for not having passed the first part of the USMLE.

The first part of the morning had gone fine. I'd found the insurance building off Boylston Street where the test was being administered and had taken my place among the other solemn students. It was during the first break that I began to sweat and feel nauseous. *This thing is really getting to me*, I thought. My head throbbed. My neck was stiff. When they let us out for lunch, I barely made it to the restroom before heaving my insides out. By the time the break was over, I had a raging fever. The afternoon stretched before me. I had no alternative but to keep going.

I don't know how I made it through the rest of the exam. I don't know how I got home without falling down in the street. Riding on the T, I still thought I was just exhausted and nervous. When I reached our apartment, I collapsed on the bed, but every few minutes I had to drag myself up and stagger to the bathroom to vomit. When Lynne came home, I asked her to give me a shot of Tigan IM to stop the vomiting; I remember steadying myself against the fireplace while she jabbed the needle in my right cheek. It didn't help. Finally I called one of my colleagues, another resident.

"What should I do?" I asked him.

"You'd better get yourself to the emergency room," he said.

Once at the hospital, one of the attendings I worked with, Mike Scott, took one look at me and told me he was going to do a lumbar puncture. I lay there like a rag, so sick by now that I hardly felt him stick the needle in the small of my back. Mike ran out and came back a short time later with the results.

"You're not going anywhere," he said. You've got full-blown meningitis."

I spent the next week in the hospital with IVs stuck in my arm. When I learned I'd passed the test, I sensed I'd not only mastered the material but also vaulted to some new level of fortitude in the process. I would not have believed I could do it until I was put to the test.

At other times during the flight, the recent scenes played themselves out in my head. I'd known my new client was royalty, but witnessing the

Saudis' extravagant life style up close was still a revelation. I'd traveled far since the days of the trailer park and the Shed—of sleeping on a cushion on the floor of a tiny apartment my first year at ASU and living off oatmeal and canned peas—but the wealth I'd seen was so incongruous with my own experience. I doubted the Saudis could begin to imagine where their new physician had come from.

And having been chosen—it was astounding. How did I get from the hardscrabble immigrant life to the one I now enjoyed, where royalty flew me to Paris and put me up in a five-star hotel? There had been a time in my youth when I wouldn't let my friend's mother drop me off at home for fear of the reaction our house would provoke. And now I'd just crapped in a golden toilet.

The house incident had been my junior or senior year in high school. My family had moved yet again. This time it was all the way up to where Phoenix ran into the North Mountains in a neighborhood known as Sunnyslope. The new rental was another cement-block box like the Shed with rooms the size of prison cells and a tiny bathroom, the floor of which was a thin remove from packed dirt. It was quiet up in that part of town. The city petered out there, and the land on the other side of the mountains was still empty. At night, the shrill, rhythmic cadence of crickets drilled into my head like an engine that keeps turning over but never starts, keeping me awake till early morning. I don't think I invited any of my friends to that house. And when the mother of one of them—Mike John-

The "Shed" where my family lived in the early 1960s. My older brother Gunther and I were proud of our VW Bugs. In those days I could not have imagined the extravagant wealth of some of my future patients.

son, a fellow student athlete—offered to drive me home, I had her drop me off up the road. I'd told myself I was not ashamed of my family's circumstances. But I couldn't be sure how others would judge us.

If the Saudis' experience was outside of the universe I'd lived in, I was sure mine was equally unimaginable to them.

On Monday I was back at Barrow. The preparations had kicked into full gear and I was satisfied that everything was ready. I went home around five o'clock, looking forward to resting and seeing Lynne and the children. My stomach growled—I usually skipped a sit-down lunch so I could make it home for soccer practice or dinner—and the aromas of a home-cooked meal wafted out from the kitchen. Around six o'clock the phone rang. It was Dr. Khouri.

"Good evening Dr. Sonntag," he said. You will be happy to know the queen's airplane is approaching. You can meet her in the private terminal on the west side of the airport."

"Oh. Well, I don't think I need to be there," I said. Everything's arranged at the hospital. The queen will have the entire seventh floor. The political and legal details have been worked out. Representatives from the embassies are prepared. Everything is arranged."

A short silence and then, "Her highness wants you to meet her at the airport. We think that is best."

"I assure you the queen will be well taken care of," I said, and wished him a good evening before hanging up.

Ten minutes later, the phone rang again. It was the doctor. "Hello Dr. Sonntag," he said. "About the airport. We greatly appreciate your cooperation. As the queen's surgeon, you will want to supervise her trip to the hospital."

Again, I politely objected, and again, after a few moments, my phone rang for a third time. The penny dropped; I realized that the doctor would keep calling until I agreed to meet the queen at the airport. Obstinate repetition was a tactic I would come to know well.

At seven-thirty I was dutifully waiting on the tarmac at Sky Harbor when two Saudi 747s landed and taxied to the gate of the executive terminal reserved for private jets. A small army of doctors, family, and attendants descended the steps and milled about with other visitors who had flown in earlier from Los Angeles. I walked out to greet the queen, realizing belatedly that she would not be coming down the stairs. I heard a mechanism whine and the cargo hatch opened. At last she appeared, riding down with her attendants on the baggage lift. There was a flurry of activity during which the now-familiar doctor found me. While the queen was eased into a kind of combination limousine ambulance, the doctor motioned me into another limousine, one of an entourage of a dozen or more luxury cars that followed the queen's to St. Joe's.

Thankfully the ride was short. As I knew they would be, hospital staff and security were waiting to receive her. They transported her up to her room and got her into bed, a process apparently requiring the oversight of her neurosurgeon. This last step accomplished, I was at last allowed to leave. Driving home, foreboding nagged at me. *This is going to be more than I bargained for*, I thought, *much more.*

CHAPTER 16

SERVANT TO A QUEEN

Sweet are the uses of adversity which, like the toad, ugly and venomous,
wears yet a precious jewel in his head.
WILLIAM SHAKESPEARE

AS I HAD INDICATED when Dr. Khouri called me at home, the entire seventh floor of St. Joseph's Hospital, which in normal circumstances housed about twenty patients, had now been given over to the royal delegation. The queen's four physicians occupied some of the rooms, and members of the royal family took up the others. Then there was her security. She was like a pebble thrown into a pond, with a Saudi security team of about thirty men forming ever-widening circles around her. Local security officers, including the Phoenix police, added to the lock-down effect.

Tuesday we ran tests on her. Wednesday—the day of surgery—opened with a circus atmosphere. Guards stood watch immediately inside and outside the operating theater, eyeing the hospital staff and talking into their lapels. All four of the queen's physicians squeezed into the OR, creating a suffocating effect. It took all my powers of concentration to block out the hoopla and focus on the problem.

Robert assisted. The queen was anesthetized, the breathing tube inserted down her throat, her neck swabbed with a sterile solution. I made a 5-cm lateral incision on the right side of her neck. Her skin was papery, the muscle immediately underlying it easily separated with my fingers. Robert put in a retractor and I went deeper, pressing the pulsing carotid to one side and retracting the esophagus and trachea to the opposite, medial side. Another retractor went in to hold this plumbing out of the way. With a clear window, and using the scope, I dissected the tissue from the spine, exposing the cervical column from C2-C7.

The degeneration was striking: The vertebrae were misshapen and porous, like pieces of coral, sea bleached and eaten away by salt. Knobby bone spurs constricted the space in the vertebrae through which the spinal cord passed. The thickened bone had also encroached on the foramen—the space on either side where the nerve roots exit the spine—closing in on the nerves like jaws. This assault on the nerves and spinal cord was exacerbated by the herniated discs, which had compressed down into brittle, yellow-red pads the density of decayed teeth.

The queen's physicians pressed in, sucking up the oxygen. I half expected my new Saudi colleague, Dr. Khouri, to offer a comment, but he remained silent. Robert and I spoke little, precisely coordinating our steps like two parts of the same machine.

With a micro Kerrison punch and a microcurette, I pried out the four discs; they offered little resistance. Next, alternating a diamond drill and a microcurrette, I shaved and scraped down the bone spurs—like a dentist reducing the stub of a tooth. This process took some time, as there were half a dozen of the growths, but once I had finished, I moved to the hip and carved some bone from the curvature at the top. From this I shaped four roughly circular plugs. "Pull on the head a little bit," I said to the anesthesiologist." He did and the disc space opened. Using forceps I slotted each disc into the vacated disc space, then placed the blunt edge of a small footplate against the disc and tapped the other end to pack it in securely. Now for the fusion.

By 1994 I had done dozens of surgeries using plates and screws, achieving mostly good outcomes. This time I selected a long Synthes plate to cover the vertebrae from C2 to C7. Once secured with two screws at the top and two at the bottom, the plate would hold the bone plugs in position, while immobilizing the spine and facilitating the fusion. The queen would live with the plate and the resulting immobility for the rest of her life. But she had already lost much of her range of motion. At least now, after the reconstruction, she would have a relatively normal curvature of her cervical spine.

We were three hours into the surgery. Throughout the previous steps, the one concern I had was the weakness of the queen's bones; they seemed a little soft. But when I placed the plate over the vertebrae and twisted in the screws, the bite felt tight. I was satisfied and we closed her up.

The queen was very sick afterwards. Per the family's demand, a neurosurgeon was posted to her floor at all times. I checked on her everyday at least once, if not two or three times. Just getting to her room was like navigating an obstacle course in a hot house, since many heads of state had sent mammoth floral arrangements that the hospital staff had done their best to accommodate. I figured two weeks would be about right for her to stabilize enough to be discharged.

At the end of the first week, I ordered a postop MRI to make sure the fusion was progressing well. The Saudis were anxious to return to their country, and I was counting the days until I could get back to my normal hectic schedule. The royal doctors accompanied me to Radiology. When I viewed the scan, I wanted to drop into a hole.

"Is there a problem?" Dr. Khouri said.

I took a deep breath. "I think the plate has come loose."

"Are you sure? How can you tell?"

"The MRI scan isn't that clear. A CT scan will confirm it."

I put a rush on the CT scan. Sure enough the plate had pulled out at the bottom, taking with it a couple of the bone plugs.

The doctors stared at the images. They stared at me. Their alarm sparked like static electricity.

"What does this mean Doctor?" Dr. Khouri said.

"It means I'll have to take the queen back into surgery and take the plate out."

"And then what?"

"Her bone won't hold a new plate. I'll have to use a halo."

The doctors erupted into a staccato exchange in Arabic. No doubt they already felt the ire of the family falling on their heads. The physicians had played a part in in the decision to select me, and now things hadn't worked out so well. Another hour passed with a detailed description of the halo brace and the complications it might cause.

Later I repeated the painful process with the family. I met individually with Prince Number One and two or three other princes and princesses. They wanted to know every detail. I explained over and over. They did not hide the fact that I had caused them tremendous distress.

The discharge, scheduled for a week out, was cancelled. I took the queen back into surgery a few days later. I pried the plate off. Now I could see just how soft one of the vertebral bodies was and took the whole thing out. I replaced the plugs and reconstructed her spine with new bone taken from her hip, supplementing that with more from the bone bank. Then I decompressed the spinal cord again, fitted her with a halo and vest, and prescribed bed rest. Later I visited her. Though I could never get an accurate take on her, having to communicate always through an interpreter, she accepted the news with grace. In the end, she seemed more reasonable than any of them.

The queen was in the hospital 107 days. During that time, every ailment known to humankind homed in on her like a smart bomb, and every time she got sick, the physicians brought in consultants. She couldn't sleep? They called an amnesia consultant. She got a gastrointestinal infection? They brought in a gastroenterologist. She had difficulty swallowing? Time to call an ENT consultant. That would have been okay by me, except that as the physician in charge of her care, I had to meet with every one of these consultants (often two or three at a time) each time they were called in. As Prince Number One put it, she had come here to see me, and that made me responsible for everything.

The family had already appropriated the whole floor walls had been repainted and furniture, chandeliers and Persian carpets installed but now they created even more luxurious living quarters for the queen and her entourage. They even rigged up some sort of system for reception of two or three Arabic television stations.

Meanwhile, other than reporting her arrival in town, the local press kept the lid on the story. The appropriate government agencies were aware of the royalty in our midst as, judging from the monstrous floral arrangements, were state departments all over the world. But at St. Joe's, anyone involved in it kept the curtains drawn. It was a total blackout.

By now I knew what I had gotten into. I was trying to keep up with my other surgeries, not to mention my private life, but it was almost impossible to find the time to do what I needed to do. If the queen hiccoughed, my phone rang or some envoy came looking for me. Prince Number One even paged me in the operating room: "You've got to come to see my mother immediately," he said when I got back to him, immune to the protocols normal mortals observe.

Travel, which was a big part of my life, became an activity akin to passing through an Israeli checkpoint. One case in particular caused a rift with the prince. He caught me in the hallway as I was leaving his mother's room, Dr. Khouri by his side.

"Ahh, Dr. Sonntag. The prince would like a word please."

"Certainly," I said, glancing at my watch. It was early on a Friday morning and I was due in rounds in ten minutes.

"We understand you will travel next week."

"Yes," I said, "that's right."

"Where are you going? Some far place?"

"I have to attend a conference in Australia," I said.

The prince bolted backward, as if dodging a blow. By this time I had discovered that he knew more English than he let on, but when he was agitated, he always spoke through Dr. Khouri. His black eyebrows pitched downward. He spoke rapidly to his interpreter.

"His highness would like to know the length of your trip."

"A week," I said. I was about to explain that, being the guest speaker, the conference had been built around my presentation and that I regretted the necessary absence, but before I could get another word out, the prince himself sputtered a response.

"No, you cannot be away so long," he said.

"I'm very sorry," I said. "It can't be helped. You can be assured that the queen will be well taken care of. My associates are capable of handling any issues that arise."

"No, no," his highness began, before spouting another stream of Arabic to Dr. Khouri.

"His highness says that a week is out of the question."

"Australia is very far away," I said.

"Simply out of the question."

It was only after I promised to stay only three days that the prince relented. So, I spent seventeen hours getting Down Under, ten hours meeting with my Australian colleagues, and seventeen hours flying back. The three-day express trip halfway around the world.

Finally, the queen was well enough to be discharged. That didn't mean she was well enough to fly home, though, so the family identified five or six huge houses in a part of town known as Paradise Valley (the Beverly Hills of Phoenix), moved the owners into a swanky resort called The Phoenician for a year and a half, and moved the queen and her entourage in. They were spectacular homes, but what recommended them above other considerations was that they were situated spitting distance from a paramedic station.

As it turned out, that precaution with the location did not work out. Some months later, when Prince Number One saw a negative news story about the response time of paramedics, he grew alarmed and ended up renting yet another house in which to install a private crew of paramedics—six in all, two per eight-hour shift for 'round the clock coverage. To be safe, he leased or otherwise acquired two private ambulances. At some point, when the portable X-ray service we provided proved too slow for the prince's liking, he also bought an X-ray machine. The prince waxed apoplectic at any lapses in his mother's care.

I would like to say that the queen's discharge made my life easier, but it didn't. Every Saturday a support team meeting was scheduled to plan out her care for the coming week, and every day I was in town, I was required to visit the queen at her temporary palace. The best time to fit that in was on my way home, and since the queen (and the members of her retinue who were around that day) waited to eat until I had arrived to share the meal, those visits took at least an hour. The only break I got at all was that the associates, sympathizing with my near servitude, relieved me from taking night calls.

In the winter of 1995, Lynne and I went skiing in Jackson Hole. Guests of the Spetzlers, we were lying in bed in the guest room, winding down at the end of an invigorating day on the slopes. We were both reading.

"Listen to this," I said to Lynne. "In Saudi Arabia if a man wants to divorce his wife, all he has to do is say: 'I divorce you; I divorce you; I divorce you,' three times."

Lynne didn't say anything, though I heard her expel air through her nose. I returned to my reading. A moment later I found another fascinating fact I wanted to share.

"Did you know that Saudi men never ask another man how his wife is. It's disrespectful. You can only ask about his family."

Lynne slapped her book down on the bed.

"Volker, I don't want to hear anything more about the Saudis. I am sick of the Saudis. We came on this trip so you could get away from the Saudis and what do you do? You bring along one of your books about them."

Her words startled me. It was true that I'd become fascinated with my demanding clients' country. I had six or seven books about Saudi Arabian culture and history lying on my night stand at home.

Lynne sat up. "These people are consuming your life, Volker. They own every free minute you have. You're visiting her on Saturdays. You're eating dinner with them. We're trying to have a normal family life and part of our culture is having meals together. You're gone so early in the morning, and dinner's the only time the kids have to see you."

I put my book down. Lynne's words were a hard rap on my skull. I hadn't been thinking clearly about the situation. Like so many, I had been beguiled by the royal family's unassailable belief in their rightful power. "You're right," I said. "It is too much. It's been what, over a year now? You must be tired of the situation."

"I am . . . thoroughly tired of it," she said.

I nodded. "You know, I am tired of it too."

That conversation took place in the privacy of the Spetzlers' home nearly a thousand miles away from Phoenix. But a couple of days after our return, one of the princes approached me bearing a lacquered metal box the size of a home printer. Fixed to the lid were a heavily wrought, bell-bedecked necklace and bangle and a heavy ring with a large, oval stone. "For your wife," he said. "With compliments from the queen."

I took it home to Lynne. She looked at the gift askance before opening it. "What? Are they bugging you now?" she said.

"I don't know," I said. "Maybe it's just a coincidence."

She lifted the lid. Inside lay a richly woven Kaftan and, nestled in its folds, a necklace of amethyst beads interspersed with seven fan-shaped gold medallions.

"Really? A coincidence?" she said fingering the Kaftan. "Well, it certainly is interesting timing. They can obviously do whatever they want."

Lynne had a point, but I wasn't finished. She could see I was hedging. "What?" she said.

"I know how you feel Lynne, but I think a short visit from you would be in order, to thank the queen."

Lynne closed her eyes and sighed. "All right Volker. I'll go with you on one of your evening visits. But that's all."

Lynne stood by her word. I think she softened a little when she saw how the queen's family so thoroughly managed her affairs. Still, as we drove down the long driveway on our way out, Lynne turned to me and said, "That's it then, Volker, right? I don't want to hear another word about that shrunken up little raisin of a queen or her family."

Lynne didn't mention her frustration to me again. This despite the fact that over the rest of the queen's treatment, I was also frequently asked to "take a look at" (I don't think this is inaccurate) every one of her relatives living in Phoenix at that time: all the sons, the one daughter, the sons' children, the daughter's children. It was phenomenal to find so many spine complaints in one family.

So, like a freight train with a hundred cars slipping through Gallup New Mexico, the situation went on and on. Every three months, we loaded the queen into her car or the ambulance, fell in with the caravan of three or four cars that accompanied her, and took her to Barrow for X-rays or another CT scan to see how the fusion was progressing. At times she seemed to be making good progress, but then one of her maladies would strike or she would have a setback of some sort and her rehab would be postponed.

One time, she just about had a foot on the plane to Riyadh. Everything was set for her departure. The family had even ordered a couple of Lincoln Continentals from Detroit specially fitted with a back seat that slid out like a slice of toast at the press of a button. Sadly, just as her departure date approached, Timothy McVeigh blew up the federal building in Oklahoma City, conjuring up fears of a terrorist attack and sending the queen back to her rental. Another time she was ready to go again when her husband, the king, had a stroke, making the political situation at home unstable and forcing her to stay put.

It did end eventually, but like everything else connected to the drama, the curtain fell slowly over the scene. One day, just short of two years after she had landed at Sky Harbor, her caravan finally retraced its steps to the airport. I rode with her in the Lincoln, along with her own doctors, out to where a jet stood ready to take her to Los Angeles. They trundled her onto a huge throne-like divan and wheeled her on board, after which the rest of us boarded with our own two feet. I flew over with them, checked into the Beverly Hills Hotel overnight, and made sure she settled into the mansion they had located for her without any setbacks. It was not too far, I heard, from Elizabeth Taylor's house.

I visited the queen six or eight times over the next year. She developed problems in her lower back and suffered recurrent illnesses. She got more infections and grew even frailer. She never did return home to Saudi Arabia. I heard she passed away in LA in 1999.

After it ended, the visit of the Saudi family seemed like one of those monsoon storms that blow up from the Gulf of Mexico every July. They dump a load of dust over everything and spit just enough rain to make a thick paste of it on cars and patio furniture. They cause a few power outages and flash floods. They get everyone excited for a spell. And then they blow out again and everything gets back to normal. My life did not exactly settle down after the queen left, but at least I stopped getting paged in the OR and got to eat dinner at home.

As for the effect of this visitation on Phoenix, much was written about the event after the fact, and there were many estimates on how much revenue the city gained during the twenty-two months of her stay. Among some of the expenses that the Arizona Republic reported, in July of 1996, in a column with the headline "Saudis splurged during Valley stay," the following figures emerged: Number of rooms appropriated at BNI: 28, for more than 3 months; number of rooms rented at the exclusive Arizona Biltmore for two years straight: 30; number of rooms rented at the Phoenician for as long as two years: 80; number of houses ranging in size from 5000 square feet to 7000 square feet, rented in Paradise Valley, for $30,000 to $50,000 per month: 6; number of vehicles intermittently rented at the same time for twenty-four hour sedan service: 50; number of five-gallon buckets of honey bought every six weeks from a local company: 24.

Of course those of us involved with the affair had a front seat to the inexhaustible flow of money; word had it at the hospital that The Phoenician built a third golf course just from the profits earned off Prince Number One's yearlong stay in the presidential suite. And when shopping at Neiman Marcus one day, Lynne heard from a sales associate that the Saudi women regularly came in and cleared every handbag off the shelf. No doubt there were other jaw-dropping expenses, and no one could ever tally it all up. Still, though Barrow never got a gift from the family, the general estimate is that the queen's stay in Phoenix, with the attendant support of her entourage and all the ballyhoo that accompanied that, had the same economic impact on the city as hosting the Super Bowl would have had.

And we didn't even have to build a new stadium.

CHAPTER 17

AUF WIEDERSEHEN

*But there's a story behind everything. How a picture got on a wall.
How a scar got on your face. Sometimes the stories are simple, and sometimes
they are hard and heartbreaking. But behind all your stories is always your
mother's story, because hers is where yours begin.*
MITCH ALBOM

THE STRANGE THING about the queen's absence was the hole it left in my life. She really was a very gentle woman, and for two years she had been a big part of my existence. I had grown used to her and her family, and I had enjoyed learning about Saudi history and customs—despite Lynne's well-founded exasperation. I looked back fondly on my conversations with Prince Number One, on his propensity to speak in English when he was pleased with me and to use a translator when he was angry; on his ubiquitous cigarettes and cups of tea. I remember how passionately he spoke about his belief in the innocence of O. J. Simpson. More than anything, I respected his devotion to his mother. I think he ended up giving himself a heart attack over it all.

I didn't just learn about the Saudis though. It was during this time, when I was pushed to the max, that I learned how to keep my family central while juggling intense professional duties. Our three children—Alissa, Christopher and Stephen—were becoming involved in more activities, and it was a priority for me to continue coaching their sports teams and spending time with them. Somehow I managed to do that amidst the ceaseless phone calls and summonses and emergencies, not to mention my other patients and administrative duties.

The queen was sensitive to my concerns about family. She had raised five sons and a daughter of her own. When I brought my youngest son Stephen by to visit her, she made a real fuss over him, and when it was Chris's turn, she gave him a Cartier watch, of all things. Then there had been that effort to make Lynne okay with the time they demanded of me. It was not that the queen's gift to Lynne had smoothed the resentments that were simmering, but I appreciated the recognition of the sacrifice Lynne was making as my wife. I thought it was a nice touch.

It was a nice touch, too, when the queen presented a platinum watch to me, a present from the king, she said. It was not the monetary value of that watch that impressed me but the way the queen glowed with pride when she gave it to me, the way she pointed out the king's likeness on the face. Her tone conveyed the same warmth and pride in her husband as any fond wife's would.

So, I think that recognition of the importance both sides placed on family helped ease a difficult situation; it helped them to understand I did have a limit, albeit one they stretched, and it helped me to remember that my demanding patient was not just a queen, she was a mother, one who was very sick.

I knew what it was to be the son of a mother whose health was failing. I knew what it was to sense time collapsing and inevitable loss approaching. I had been in the prince's position a decade earlier. Though I have no idea what kind of relationship the prince and his mother had, when I found that my own mother was terminally ill, what struck me was the sense of things unsaid, of time accelerating to rob me of any chance to rectify that silence.

My mother lived a dozen years after my father's death in 1974. She retired. She moved to a condo near where Lynne and I lived. She traveled and spent time with friends. She took great pleasure in seeing the next generation of Sonntags come along. After the early years of privation and struggle, I think her last two decades were full and satisfying. Still, I didn't know much more about her than I ever had, and nothing in our interactions led me to believe that this would change. I had not been alone with her since the two of us had traveled to Germany to deliver my father's ashes to a cemetery in Timmendorfer Strand.

Lynne and I were juggling the usual domestic and professional balls during those years. Our family was growing—Alissa was five years old that year, and Christopher was three—and we had our niece Tatiana, Gunther's daughter, living with us in our guesthouse. With my career taking off and Lynne heading up the household—she had given up her profession as a nurse pediatrician a year or so before—we had all we could do to keep apace with our responsibilities. Maybe that is why we did not notice anything unusual about my mother until, in December,

Gunther's wife, Lettie, came to visit. Before she left, she pulled us aside and said, "What's wrong with Mutti?"

I think my mother sensed that the Christmas of 1985 would be her last, and she wanted everything to be perfect. As we had done during her later years, we celebrated Christmas Eve at Lynne's and my house, incorporating all those German traditions that had brightened our home since the early days in Arizona, when my parents had barely had enough money to send in the monthly payment for the trip over from Germany. This year, with our large, extended family gathered all around, those German Christmases of the past seemed real again—the candles on the trees, the dining table laid out with fish and cheese and jam and herring *Rollmöpse*, the laughter of the children.

That evening, we went to church and prayed and sang together, as we had back in Bad Hersfeld before my father got sick. Even as we drove down the wide Phoenix streets, past the palm trees and the cement block houses and the strip malls, scenes of those days in Germany rose up in my mind.

I remembered how we had bundled up against the German winter and walked through the darkening streets to the city center, how the snow on the roofs of the old buildings caught the fading light of the day as the slopes of the mountains surrounding the old town faded into silhouettes. I remembered the Gothic tower of the *Stadkirche*, rising up over the ancient timber-frame houses with their red doors and geometric patterns. We were in the desert now, but the spirit of *Weihnachten* was all around.

After church, we came home and sang again the German versions of "Silent Night" and "Oh come All Ye Faithful"—"*Stille Nacht*" and "*Herbei, o ihr Glaubigen*." I read the story of the Nativity from the same Bible I had studied all those years ago. We ate and laughed, and my mother and I and Tatiana spoke yet again about our ties to Germany. The family had heard it many times before Lynne's parents and her brothers and their families but it had become part of the ritual.

Joy, religion, tradition, togetherness—that Christmas everything was as it should be.

It was only when the glow of the holiday had passed that my mother told us about the blood in her urine.

"Why didn't you tell us earlier," I asked her. "Why did you wait so long?"

"Christmas was coming," she said. "That came first. Now that it has passed, we can take care of this problem."

Soon after that conversation, she had an exploratory laparotomy at Saint Joe's. I paced around in the waiting room, afraid of what they would find when they opened up her abdomen, hoping, against reason, that there was some other explanation for the bleeding than the one I suspected. There was not. She was riddled with tumors. All the doctor could do was close her up.

When she surfaced from the anesthesia, I was by her side.

"What did they find?" she asked.

"It's nephroma, Mutti," I told her, "a tumor in your kidney." I waited a moment before continuing. "It has spread to the abdominal cavity and to the other organs."

"How will you cure it?" she asked.

I took a breath. "There is no cure," I said, gently, "and no treatment."

She was silent. For the first time, we looked at each another with tears in our eyes. Later, when I left her, I could hardly say goodbye. When I did, she corrected me.

"Not 'goodbye,' Volker. *Auf Wiedersehen*, till we meet again."

My mother emerged from surgery frail and depleted. Lynne and I moved her from the condo into our guesthouse. Still, where her body failed, her mental power and will were rock-solid, as was her pride. While my mother resigned herself to letting Lynne nurse her and bathe her, to let Lynne see her in her weakest state, she refused to let me see her like that. I had to smile at that reversal. My mother had been a formidable mother-in-law, a matriarch who had told Lynne shortly after meeting her that she didn't like her laugh and that she should change it. Now she entrusted herself to Lynne.

Many visitors came. On those days, she refused to take any pain medication. It was unthinkable to her to be sleepy or incoherent when so little time was left with those she loved. Some visitors, we knew, she would never see again, but always she parted from them with the same words: *Auf Wiedersehen*. Struggling to control her emotions when she took their hands, she always brightened up once the visitor had left. Each day was a small gift she looked forward to.

It was after her sister, Tante Gu, came to visit that my mother seemed to accept the inevitable. Lynne and I had been away. When we came into the guest house upon our return, she looked me in the eye and, in a matter of fact way, as if she were talking about vitamins or diet pills (but also with a touch of anger, as if we had been trying to dupe her) said, "These pills I have been taking aren't going to help me, are they?"

"No," I replied. "They are not going to help you."

She turned to Lynne and said, "Light me a cigarette will you?" and puffed on it while Lynne held it.

Easter arrived. Weak as she was, my mother wasn't done with the world. As she had at Christmas, she orchestrated the festivities and insisted that Lynne make Easter baskets for every member of the family. Those preparations took me back again. I remembered how she and my father and I had met up for Easter sunrise services the years I was at ASU. I remembered how, back in Germany and later in Phoenix, she had instilled in us a sense of mystery and ritual over the four days of Easter, how, following tradition, she had admonished us not to speak on Good Friday until she had come around to offer us a bite from an apple.

A week or so after Easter, we moved her into a hospital bed in the living room of the guesthouse. Lynne stayed with her during the day, and visitors called to say their goodbyes in the evening. Through that difficult and emotional time, she repeated the same phrase to all the loved ones who came and went—"*Auf Wiedersehen*. Till we meet again, if the Lord grants it so."

The time of vigil came. Each day, we gathered in the guesthouse for dinner and then passed the evening with her until it was time to go to bed. One evening after dinner, my mother gave a final gift to my brothers and me.

Lynne had left to give the children their baths, leaving Gunther, Rüdiger, and me alone with her. I couldn't remember the last time the three of us had been alone with my mother. Rüdiger lived in town now, but Gunther had flown out from Greece for what we knew would be my mother's last Easter with us, and then delayed his return.

The heat had come early that year, but now it had lifted. We were passing the time, as people do in such situations. Suddenly, like a sleepy person who gets a second wind, my mother sat up in bed. Her face came alive and her eyes were bright.

"I have something to tell you," she said, "something about the past."

That happened just a few days before the end. The stories my mother shared with us after forty years of silence were wandering around in my head trying to find a home among the few facts I had known. She was not done with me yet, though. The night before she died, I was alone in the room with her. For the second time, my mother pulled a card out of her sleeve. She turned to me, her voice a whisper.

"*Ich liebe dich*, Volker," she said. "I love you, Volker."

"*Ich liebe dich*, Mutti," I responded. "*Ich liebe dich*."

The words came to both of us as if we had said them every night for the last forty years. In fact, it was the first time we had ever spoken them.

She raised her arms and took a ragged breath. I put my ear to her lips. "Stay healthy and strong," she said. Those were the last words she spoke directly to me. Somehow, they were perfect. Somehow, I had always known that is what she wanted for me, and I had lived by her final wish for me all my life.

The next morning when Lynne and I arrived, she was in a dreamy, twilight state. She knew we were there, though. Just when I thought I would never look into her eyes again, she opened them and looked at us. A peaceful smile played on her lips.

"Am I still here?" she said.

And then, "I will go now."

That was the last time I saw her alive. That afternoon, my mother's old friends from the trailer court days came to say goodbye. Lynne told me that my mother was awake for that, awake when the couple carted off the padded lounge chair they had brought over for her to use. "Guess she won't be needing this anymore," they said. And they were right. When Lynne returned after walking them out, my mother was gone.

It was April 24, 1986, Lynne's and my wedding anniversary.

I carried my mother's ashes back to Germany with me, aware of her presence in the overhead bin. I wasn't going to have her sent over like a parcel. She was laid to rest in Winnenden, near her mother, where her surviving sisters could tend the grave. I didn't question why she was not buried with my father in Timmendorfer Strand. He had also been laid to rest near his mother.

On the plane over to Germany, I wrote the eulogy for her. I wanted to honor my mother's indomitable spirit, her strength and her resolve, but I also wanted to acknowledge the way, at the end, she had not forgotten about love. So, I wrote about her last Christmas and her last days. I wrote about how she had told me, at the end, that she loved me.

The day of her funeral, a drizzle whispered down from the swollen clouds. Even the sky—a good German sky—couldn't quite cry. My mother's three sisters were there, one of the uncles, Gunther and I. When the time came for the eulogy, all the unexpressed emotion of my family's past welled up in me. I couldn't speak. I gave the passage I had prepared to the pastor. When he came to the words *"Ich liebe dich,"* he stammered. He seemed embarrassed. Like the German sky that couldn't cry, this German could not give voice to the words "I love you."

It didn't matter in the end. My mother and I had made our peace with the silence of all those years.

CHAPTER 18

CHIEF AND CHIEFIES

Honesty prospers in every condition of life.
FRIEDRICH SCHILLER

THE HOUR-AND-A-HALF teaching rounds I conducted each Wednesday and Friday had just begun. It was late in the day but early in the academic year, and for the sake of the newest residents, I was addressing the Glasgow Coma Scale (GCS).

The GCS is an assessment with which a doctor tests a patient's neurological condition—his alertness or level of consciousness—in three areas: the eyes, motor response, and verbal response. Introduced after I'd finished my training in the 70s, the scale sure beat the old highly subjective test where you called a fellow doctor on the phone and said: *Listen, I've got a guy here who's lethargic, semiconscious. What do you think?*

I scanned the variously fatigued, nervous, engaged, or confident faces before me—teaching rounds were always a mix of all levels of residents with the fellows and some attendings present as well—and shot a question out. "OK, in the category of the eyes, we know an obvious sign of severe brain stem injury is dilation of the pupils. Now describe the doll's-eye reflex, what it signals, and identify the system that would be affected if a patient exhibited the sign. You, Dr. X., what do you think?"

I called on a junior resident intentionally, to push him to retrieve what he had learned. I could tell before the question left my mouth, however, that the new recruit didn't know the answer. "OK," I said, "let's ask Chiefy over here," calling on a senior resident. He didn't know the answer either. "Holy Toledo," I said, "that question is not exactly part of the GCS; maybe that threw you off. Let's move it up to the chief resident."

So began another teaching session, a duty that by the early 2000s I had discharged for nearly twenty years. When I recall how I rejected the chance to go into academics under Ben Stein, fresh out of my residency, it seems fateful how I got pulled back into teaching when I started doing the board review rounds for the residents. That was at the end of the 70s, when I had been in Arizona only a year. Then in 1988, amid the controversy over neurosurgery's role in the spine, I started the spine fellowship at

Barrow. As the 90s progressed and my career became more demanding, I thought my plate was as full as it could get. But in 1995, in the middle of the queen extravaganza, Robert asked me to assume the role of Residency Program Director. There was no question of refusing the task. I resonated deeply with Robert's vision for BNI as a provider of excellent care and a premier teaching institute. Helping carry it out was an extreme privilege.

Robert's vision was fast becoming a reality. Through the 1990s, the institute expanded rapidly; it began to attract top-tier residents; more fellowships were being done; and we began to cast a much bigger shadow both nationally and internationally. If I was having successes with the spine, other doctors were hitting their own balls out of the park, chief among them Robert. Already possessing a reputation that brought in referrals from across the country on the toughest and rarest cases—even those deemed impossible—Robert had mastered and then improved upon the neurosurgical equivalent of a show-stopper, the cardiac standstill procedure, otherwise known as hypothermic arrest. To recall, that procedure involves basically stopping the heart, lowering the temperature of the brain to about sixty degrees, and draining all the blood out of the body. It leaves almost everyone awestruck, and Robert performed it repeatedly and with great success. In the early 1990s, a writer named Edward J. Sylvester wrote a book about neurosurgery called *The Healing Blade*, highlighting Robert's remarkable (some would say godlike) powers. If anything comes close to raising Lazarus from the dead, the cardiac standstill procedure is it.

Under Robert's leadership, Barrow became the busiest neurosurgical service in America. That growth propelled a number of BNI associates onto the big stage as well. Hal Rekate, the pediatric neurosurgeon who had worked on TJ's case with me, became a national figure in the 1990s. Another neurosurgeon, Curtis Dickman, who had started as a resident with us and then done a spine fellowship in Florida, did his part to put the spine even more securely on the map. And a Ph.D. biomechanical researcher named Neil Crawford rolled out great work for us, testing out new plates and techniques on cadavers and with mathematical models. There were other outstanding colleagues too, among them Nick Theodore, a resident, and Stephen Papadopoulos, my third fellow, both of whom supported the advances we were making in the field and who eventually became associates and good friends. These and other top-notch people did remarkable work at Barrow in those years.

Given our tremendous progress then, by the middle of the decade, it was clearly time to ratchet up the residency program. That was a huge job administratively, and I won't go into all the details except to say that with three new residents (later four) joining the program each year for a seven-year training period; with overseeing their education and doing six-month evaluations on each resident; with visits from the Resident Review Committee (RRC) and compliance with regulations; with attending week-long meetings at the national conferences and integrating recommended changes; with teaching rounds and the preparation that goes into them; and with such lesser but no less critical labors as mediating disputes among a highly ambitious and stressed group of individuals, my job even then was only half done.

Having said that, the heart of the residency program and my main goal as director was to teach and mentor the young doctors to the best of my ability. Towards that end, when it came to my teaching approach, I drew from my own training and experience.

By the mid 90s, I had "made it" in my career, but I had not forgotten how it felt to be a resident, much less a lowly med student on rounds, waddling along like a duck behind the attending and the residents, steeling myself for a question at the patient's bedside. I remembered the uncertainty and impotency of my first rotations in surgery, the angst inspired by the eagle-eyed scrub nurses who pounced on the med students. And I had experienced that breed of senior surgeon whose teaching method consisted of biting criticism, made all the more belittling by being done in the presence of the student's or resident's peers.

My chair at Tufts New England Medical Center, Ben Stein, had been tough like that in the early years, and there had been one occasion during the first year of my residency that had shamed me deeply.

Stein had sketched out a cross section of the brain stem on the blackboard. "Okay, here is the brain stem where the brain connects to the spinal cord," he said, turning to the residents. Then he walked over to me and thrust a piece of chalk into my hand. "Dr. Sonntag, draw the course of the fifth cranial nerve through this section of the pons."

The other residents, the attendings, and the visiting neurosurgeons turned to me, waiting for my answer. I was clueless. I had no idea where the fifth nerve went. I felt my face go hot.

Stein knew I didn't know, but he persisted. "Well, where do you think it is? What do you think it might look like?"

I knew the CN-V was the trigeminal nerve, that it was both a sensory and a motor nerve that affected feeling in the face, sinuses and mouth. But I was drawing a blank as to where it ran in relation to all the other nerves.

"Go on. Draw it on the board there," Stein said.

I drew a line. Stein snorted, grabbed the chalk from my hand, and said, "Sit down Sonntag."

However much I respected Stein and the way he pushed us to use what we had learned and to build on it, once I began teaching residents, I was certain that I would never model my own teaching strategy after his. Now that I was on the other side of the equation, I strove to be demanding but fair in my role as teacher. I thought of the residents and fellows as family, and the manner in which I treated them naturally followed: respect each individual and teach by example.

A little humor never hurt either. "Note that the lower limit on that Glasgow Coma Scale is 3," I told the group that day. "Even when a patient's brain dead, he gets points just for showing up."

"Chiefy" in the mid 1990s wearing a gag brain cap. By then I had added humor as a fifth "H" to my philosophy of the 4 Hs: hard work, hope, honesty, and humility.

Every Monday and Friday morning I held teaching rounds on the spine. This was one of the real strengths of Barrow's residency program. In those pre-digital days of the 90s, we gathered in front of a huge X-ray screen in the radiology department, the two fellows and one or two attendings joining the fifteen or twenty residents available.

People usually think of rounds as the circuit of patient visits that an attending doctor makes with med students and residents. But teaching rounds are more a group case study, a real-life application of topics on which the residents are developing their expertise. The pedagogical materials used at these sessions were (and are) letters of referral from other neurosurgeons—not just local or regional but from around the world— each one accompanied by diagnostic studies and X-rays. We got ten or twelve of those a week. Added to those cases would be any interesting spine cases that had come into Barrow and been referred to me by one of the attendings.

The resident on the spine service would have come in early and put all the films up. After the day's work, I'd walk in, maybe crack a joke or two to get things going, and we'd start. Case by case the resident would describe the patient's history and the details of the case, after which I would throw out questions. Kicking the question up the ladder of experience from junior resident to senior resident, chief resident, fellows, and attendings made the session relevant to all those in attendance no matter what their level of expertise.

For me, the supreme vehicle for passing on highly complex knowledge from teacher to student lay in the relationships I formed with the residents and fellows. That meant knowing them not only within the context of work but outside the hospital as well. For years that interaction took the form of informal volleyball or basketball games at my house on a late Friday afternoon or soccer practice on the weekends, which provided rich opportunity for getting to know the people behind their roles.

Over the years other traditions got started and stuck. In the 1990s and early 2000s we had a BNI/UCLA Olympics (before the sheer numbers forced us to limit the games to Barrow under the Sonntag and Spetzler teams). Following Robert's lead, we also instituted the "Hike from Hell," a scramble through the West Fork of Oak Creek Canyon each fall, beginning with a sleepover at Robert's place in Flagstaff and concluding with a soak in the hot tub followed by shots of scotch and a few hands of poker. Activities such as these promoted healthy athletic competition but also the kind of deep camaraderie that only comes from working hard and playing hard together.

Those young men and women became my second family, and when you think of the people you are teaching as family, the manner in which you treat them follows naturally.

It was during those years that the title "Chiefy" emerged, a moniker that reflects the egalitarian and team environment that both Robert and I fostered. It was a title that applied not only to me but also to the other associates and eventually the residents and fellows as well. It was a flexible title. Depending on the tone of voice in which it was uttered, it conveyed approval, respect, or praise—or the opposite of each of those attributes. However it was said, prefacing any comment with the word "Chiefy" imparted a sense of inclusion and a commitment to an even playing field in the work environment. I think that seemingly small practice contributed greatly to our success and our well-being at work.

Harsh criticism has always seemed unnecessary and counterproductive to me. In my role as teacher, I figured I was there to educate and influence these individuals, not to pull rank on them, not to imply, *I'm the big neurosurgeon and you're in medical school and that's the way it is. Screw you pal.* That is not only a miserable attitude to have about your students but a shortsighted approach to dealing with people who will be your colleagues and peers in the future.

Moreover, such a sorry attitude takes no account of what those people have done to get as far as they have. The neurosurgical residents and fellows were, and are, highly intelligent individuals. They wouldn't be sitting there next to you if they weren't first rate. So, when, as the teacher or mentor, I called on a person, and I realized that that individual didn't have the answer, instead of resorting to sarcasm or insults, I moved on to the next person. I figured the individual I had first asked knew that they didn't know the answer, and knew that they should have known it.

I think of teaching as mentoring, and again, I have always tried to teach by example. I figured if I worked hard, the residents would work hard. If they saw that I put my family first, well, hopefully they would do that as well. Judging from the stellar careers of many of the 55-plus residents and 36 fellows that have come through Barrow's program during my tenure as residency director, I'd say that my approach has worked well. Still, there were those that didn't make the cut. Dealing with the necessary discipline in those cases was one of the hardest things I had to do.

Ken Adams was the nicest person you could hope to meet, a real gentleman. Trim, soft-spoken, and respectful behind his thin-framed glasses, he'd been in the residency program about a year and a half. Now he stood at the door to my office.

"Come in, Ken," I said, "and close the door."

Ken did as I had directed and sat down across from me. He knew something was up. A resident was never called to the chief's office unless there was trouble.

"You know why you're here, don't you Ken," I said.

So began a conversation I only had to conduct three times over fifteen years in my capacity as resident program director. It never got any easier. But once the news of a problem got back to me through the grapevine, it was unavoidable.

On this occasion it was the chief resident who had caught me in the hallway outside my office.

"Chief, can I have a word with you? We've got a problem."

"Yeah, what is it?" I said.

"It's Adams. We were doing rounds on Tuesday and looked in on a neck case slated for surgery that day. I said to Ken, 'Hey, did you check the CT scan on this one? Is his head clear?' And he said, 'Yeah, I checked it. All clear. You can take him into surgery.'"

I nodded. "OK, then what?"

"Well we took him in. The surgery went fine. But when we got the patient into recovery, his blood pressure spiked and his heart rate dropped. It looked like he was going to code."

"Cushing reflex," I said. "Obviously more going on there than the neck."

"Right. So we looked at the scans again. Turns out he had an intracerebral blood clot. We took him back in for an emergency craniotomy. He pulled through. But there's no way Adams could have missed that clot on the CT scan."

"Did you ask him about it?"

"Yeah. At first he just repeated that he'd checked the scan. But when I pressed him, he admitted he hadn't looked at it. I don't know why he didn't just say so before we took the patient in."

I didn't say anything, but I knew what had happened. Adams was on rounds. Twelve or fourteen of his peers were standing there. He wanted to look responsible, to show that he'd done his job. What was he going to say? That he'd skipped a crucial step? He knew that if he told the truth, the chief resident would come to me, and that wouldn't look good. So he lied. Ninety percent of the time, he might have gotten away with it. But this time the scan was abnormal. He lost the gamble. Then, once called on it, he would have been in much less of a bind if he had told the truth. Now it was a huge problem.

"All right," I said. "I'll talk to him."

The meeting lasted no more than ten minutes. At this point, the protocol was clear. When it comes to patient safety, there is no minimizing, no misplaced mercy. I was not oblivious to the fact that Ken was a young doctor whose future—at least in part—rode on my evaluation. But my hands were tied.

"You know why you're here, don't you Ken," I said.

"Yes, I do."

"I understand you failed to check the CT scan before a surgery and missed a blood clot."

He swallowed. "That's right," he said.

"You know this is a major issue of patient care and a clear violation of how we do business at BNI. Your action endangered the patient's safety. Patient safety is always our primary concern."

"I know that," he said. "I'm very sorry. It won't happen again."

"I trust that it won't. But this is a major, major problem. I am going to have to put you on warning status. That means if you have another major incident during the three-month warning period, I'll have to put you on probation."

He heaved a sigh and looked me in the eye. "I understand."

He took it pretty well, stealing intense glances at me as I dictated notes on our meeting for documentation. When I had finished the recording, I printed off a report and signed it.

"This will be part of your permanent folder," I said, having him sign it after me. Then I stood up, and walked to the door.

"Good luck," I said. "Let's hope you'll be okay over the next three months."

I closed the door after him, then walked over to the window and looked across to the building opposite. I felt a tremendous letdown, as if I myself had failed in some way. I knew the stakes. It was a damn shame.

Initially things settled down. After my bi-weekly meetings with the residents, I'd ask the chief resident how Adams was doing. *He's toeing the line*, I'd hear. I'd talk with Adams himself, once or twice a week. During the warning period, however, there was another incident. I don't remember that second one, but Adams was put on probation. Then, about six months after the first meeting with him, I got wind of yet another infraction. This time the report came from the head nurse.

Virginia Prendergast was, and is, a strikingly intelligent Ph.D. with a wide smile that immediately puts people at ease. Beginning as a staff nurse in the neurosurgical intensive care unit, she worked her way up the leadership ladder, defining and developing BNI's formal neuroscience nursing educational program. At the end of the 90s, her teaching achievements and dedication to patient care came to command national and international recognition and awards. Given her passion for excellence in patient care, when she poked her head in my office one afternoon to tell me of the problem, I had no doubt that her perception of the situation was accurate. What she had to say sealed Ken Adams's fate at Barrow.

I was sitting at my desk when Virginia knocked on the open door. "Got a minute, Chief?" she said. "I think you need to be aware of something."

"Sure," I said. "What is it?"

She closed the door behind her. A deep crease cut between her eyes. "It's one of the residents, Ken Adams. He was at a patient's bedside this morning, pulling out a shunt tubing. I was passing by in the hall and I heard the patient asking him to stop. So I looked in. He kept pulling at it, even though the patient was clearly in pain. The patient finally started screaming and asking for pain meds. I really couldn't believe what I was seeing."

I wasn't surprised to hear this new problem. I was disappointed though, for Ken's sake. I hated to see anyone fail, and if there had been any way to rectify the situation in an ethical manner without formal disciplinary action, I would have taken that route. I may even have erred at times giving one or two residents the benefit of the doubt. That's what my secretary Cheri told me. "You're too soft, Boss," she'd say. "You let the boys off too easy."

But I was angry too. What Virginia reported to me was not only a serious departure from policy; it was also a shocking disregard for the patient's comfort. A brain shunt is a narrow piece of tubing that is first inserted into a fluid-filled ventricle of the brain. Able to be inserted into any orifice of the human anatomy, the tubing is then passed under the skin into another area of the body, most often the abdomen but sometimes one of the chambers of the heart or the lining of the lungs. The shunt tubing relieves pressure on the brain, a condition known as hydrocephalus, by draining the extra fluid in the brain ventricle to a different area of the body where it can be absorbed more quickly. To remove it requires extreme care.

"Unbelievable," I said. "Adams has been on probation, you know. This is strike three."

"I knew that," she said. "I can't imagine what he was thinking. Or not thinking."

"I'll take care of it," I said. "Thanks for informing me of the incident."

Ken was back in my office that same day. It baffled me how he could have let the situation get to the point it had.

"Ken, I am sorry to see you back in here. It's been reported by the nursing staff that in removing a shunt tubing today, you kept pulling on it even though the patient was in extreme pain. The patient apparently asked you to stop, and also requested pain medication, but you ignored the request and you just kept pulling. Is that what happened?" I asked him.

Ken concurred that the account was accurate.

"Ken, again, this is a matter of appropriate patient care. Did you think that was appropriate?"

"No, I guess it wasn't," he said.

"I'm afraid that because of this incident, I have to let you go."

I let that sink in for a moment. Ken said nothing in his defense. He sat erect, his eyes downcast and his mouth clamped tight.

I hated such moments.

"I'm sorry it hasn't worked out for you here at Barrow," I said. "Maybe in a program that is not as hectic as this, you might make it. At the BNI we have an extremely busy emergency department, an extremely busy service. We do more surgeries than any other institution in the country. It's a fast-paced, high-volume institution. You might consider going to a place that has a slower pace. You might consider going into neurology."

"All right, Chief," he said. "What do I do now? What about a recommendation?"

"Well, if you go to a place where you might fit in, they'll know we fired you. There's no way around that. But if it's a place that's not as high tempo, I'll talk to the program chair and tell him what I just told you."

He nodded.

"You know, it's like a marriage," I said. "You get engaged for a year or two. You get married. And then you find out it wasn't the right person. There's nothing wrong with either side. It just wasn't a fit."

I wrapped it up after that. I shook his hand and he left. That was the end of Ken's residency at BNI. But it wasn't quite the end of the matter. A sour wave or two splashed up in the aftermath.

The first one was the news that Adams had gone to Robert Spetzler and asked him to reverse my decision. I could have told Ken not to waste his time. Robert told him what he eventually told the other two residents I fired: "I stand by Dr. Sonntag's decision," he said. Robert was savvy enough to know that to do otherwise would have undermined not only my position but also the integrity and leadership of the institution itself.

A couple of months later the second wave hit shore. I received a letter from a lawyer informing me that Adams was suing me for wrongful termination. Robert received one too. I remember spending half a day answering questions in a lawyer's office across the street from St. Joe's. In the end, Adams's legal action went nowhere. Several months later another letter arrived informing me that the case was without merit and had been dropped.

That still wasn't the end of his legacy however. Years later another missive came bearing news of Ken Adams. There was no letter, just a photocopy of a certificate showing that Ken had completed a residency program at another hospital. Then, another arrived, certifying that he had passed his written board exams. Over the course of the next several years I received other letters detailing Ken's successes.

I can honestly say I was happy to hear of them.

CHAPTER 19

THE COURAGE TO CARE

Success is not final, failure is not fatal:
it is the courage to continue that counts.
WINSTON CHURCHILL

DISHONESTY IS unacceptable in any workplace, but nowhere more so than in a hospital. Still, it is there, surrounded by fellow physicians, that an individual may feel especially pressured to "doctor" the truth. After all, doctors—perhaps more so than any other professionals—are supposed to know what they are doing. Moreover, professional reputations are fragile constructions. When a doctor screws up, he knows that word of it will travel through his community of colleagues like wildfire. No doubt Ken Adams's keen awareness of this buzzing, merciless grapevine led him to take the shortcut that he did. That move was not only an ethical breach but also a failure to consider the possibly devastating consequences of his dishonest act.

I could well remember the stress and uncertainty I felt when confronted with a situation I didn't know how to control. There was one incident during the third year of my residency, when I had already moved up to chief resident and had developed some capacity to deal with the unexpected. Everything had gone fine in the OR when Stein opened up the carotid artery on a patient, cleaned it out and stitched it up. Then he left.

I was on call that night and had been monitoring the patient's condition post-op. I was in the ICU with him about two o'clock in the morning when his neck blew open, spraying blood all over and spurting like a back yard fountain. I ran and got some gauze (in those days we didn't think about pathogens or AIDS or anything like that; we always got spattered with blood) and applied pressure to the patient's neck. I buzzed for the nurse. "Get Stein on the phone, stat," I said, when she finally arrived. She dialed him, waited for him to pick up, and handed me the receiver.

"Dr. Stein. The carotid just blew on the endarterectomy you did today," I said. "What do you want me to do?"

"How's the patient?" Stein said, his voice groggy.

"The patient seems to be okay."

"Well, wait for the morning." With that, he hung up.

Wait for the morning? I thought. *And do what?*

So, I waited until morning. I sat there the rest of the night, swapping out gauze squares and keeping pressure on the patient's neck. Around seven o'clock Stein came in and took the patient back to the OR. That was that. No further discussion. I didn't question him; that was not something you did. If I had known him better, maybe I would have asked him if that is what he had intended me to do, but that was long before I got on those kind of terms with him.

The point is that residents are bound to encounter on a regular basis situations that will choke them with anxiety. The only way I have found to head off such situations at the pass is to stifle any feelings of stupidity or foolishness and immediately own up to my uncertainty or mistakes. Kicking them down the road will only make things worse. Honesty is the only path.

I never lied about a mistake, but I certainly felt stress of this sort many times; it comes with the high stakes territory. But there are variables that can heighten the anxiety, the most salient being a bad outcome. At the height of my career, I had my share of these cases. All surgeons live with them, and many of us probably second-guess them ad infinitum.

One case I had involved a male patient in his early thirties who was already severely ill and had a very tight spine on which we did a decompression. It looked clear cut when we opened him up, and the operation went without a hitch. Later when he woke up in the ICU with no movement in his arms and legs, we were rattled. We examined him, went over our notes on the surgery, and did an emergency MRI scan. Still we could find nothing that should have caused his condition. No blood clot on the spine. No evidence of low blood pressure. No recognition after the fact of an "oops moment" during the surgery. The patient did improve, but only to the baseline at which he had come to us.

Then there was a pedicle screw case in which the patient woke up with weakness in the foot—so called "foot drop." We took him back in and found that the pedicle screw had been misplaced and was sitting near the nerve. We readjusted it, but the patient never regained normal feeling in the foot.

That last condition—foot drop—is a not uncommon side effect of lumbar spine surgery, so I had several cases of this sort. A memorable one was a courtly Las Vegas businessman in his 60s named Roger LaMotte who was referred to me with a very bad case of stenosis and curvature of the spine. I had done dozens of pedicle screw cases by then, and as was standard, I scheduled a complex lumbar spine decompression and fusion with pedicle screws.

I knew it would be challenging, but the first sign of real trouble came when we got to the foramen. The foramen is the space on either side of the vertebra, in the back, through which a nerve root branches off and exits the spine. As with the central opening of the vertebra—the spinal canal through which (at the lumbar level) the nerve roots, or cauda equina, run—normally there is plenty of room for the nerve roots to exit the spine. However, Mr. LaMotte's spinal stenosis caused degenerative changes such as bone spurs that not only constricted the space for the nerves in the spinal canal; they also deformed the foramen, walling in the space where the nerve root exited.

My fellow—Paul Marcotte I think it was—siphoned off the blood as I labored under the microscope to decompress and fuse the patient's L5 vertebra. I worked for three or four hours, cutting and shaping bone, replacing discs with bone plugs, realigning the spine and fixing the screws. It was all going reasonably well for such a complex surgery. But when I reached the foramen, I felt some trepidation. The bone had almost completely encased the nerve root, strangling it between fingers of bone where it would normally exit. It was going to be hell to free it up. Because when the nerve is really squeezed like this by the bone, any instrument you put in there to get rid of the bone is going to compromise the nerve as well.

"Boy this is a real humdinger," I said to Paul. "We're going to have to use a # 1 microcurrette to free the nerve up, that and a 1-mm Kerrison."

I worked for another hour on it, painstakingly chipping bone away with the smallest instruments possible. In the end I had to shave some remaining bone from the top of the foramen with a diamond burr. When we closed him up, I was satisfied that the nerve root was freed from the vise of the distorted bone, but uneasy about the damage that may have been done to it in the process.

My trepidation was validated when I examined Mr. LaMotte in the recovery room after the operation. I ran my hands over his legs, examining each muscle in detail. Then I gave him the test.

"All right. Move your foot," I said.

He wiggled his right foot, flexed and extended it.

"That looks good," I said. "Now try the left foot."

He extended the left foot, pushing down as on a gas pedal, but when he tried to bring his foot back up to bend at the usual standing angle—i.e., in a dorsiflex—it hung pale and inert, like a dead fish.

"It looks like you have a weaker foot on this side," I said. "It's not extending upward as well as before surgery."

"What's causing that?" Mr. LaMotte said.

"Well, given the deterioration of your lowest lumbar vertebra and the top sacral vertebra, it's not unusual" I said. "The classic foot drop—when the foot can't dorsiflex and drops down—comes from nerve damage at L5. That's the nerve that keeps your foot up. In your case, the nerve at L5 was severely pinched. There was considerable tightness in the foramen. That will cause problems moving your foot up. We'll run another CT scan to make sure everything looks okay."

We took the postop CT scan. The L5 and the S1 screws were properly placed and holding. Still, those screws were near the nerve root, and I knew that freeing up the nerve might have traumatized it.

"Things looked okay on the scan," I said when I met with Mr. La-Motte again. "That nerve was very tight at the time of surgery, but you do have some movement in the muscles enervated by the nerve. With time and rehab, the chances are overwhelming that those muscles will get stronger."

Fortunately, that's exactly what happened, but Roger LaMotte's case typifies the touch-and-go nature of surgery. And whenever such problems did not resolve themselves, you felt as if you had been thrust under a hot, harsh spotlight. Your day and week just went down the tubes. You would reflect on it from every angle you could in hindsight, and still find no logical reason for what had gone wrong.

Not that you gave preference to one patient over another, but when the case involved a well-known person, your anxiety would shoot up like a Roman candle. When a former governor of Arizona also experienced increased weakness in one foot after her surgery, I could not stop imagining her limping around and being asked, *What happened?* It wasn't even primarily the public you worried about. Even in a routine surgery (and none are really routine; there will always be a surprise) when you are visible in the community and you are working on a high-profile member of that community, news of any failure will spread like wild fire through the neurosurgical circles. Fortunately, the governor also made a complete recovery within a couple of weeks.

Happily, I did not have many such failures, but there are always cases that keep you awake at night. Among these is the difficult patient.

Not long after I watched Mr. LaMotte drag his foot out the door upon his release, another situation arose. I'd left for the day, glad to leave the misery behind for a few hours. No sooner than I'd thrown my keys down on the counter at home, the phone rang. It was a nurse from the recovery room at St. Joe's. "Doctor, you've got to come back up here," she said. "That nurse from Scottsdale Memorial, the one you operated on today, won't stop screaming."

The patient was one on whom I'd performed a common neurosurgical spine procedure—a decompressive lumbar laminectomy for spinal stenosis. It was an operation I'd done hundreds of times. Again, there had been no "oops" moment, nothing out of the ordinary...until she woke up. Then the screaming started.

I was putting the phone down when Lynne walked in wiping her hands on a towel.

"I was just going to set the table for dinner. Is there a problem?" she said.

"I've got to go back to the hospital."

She pressed her lips together and sighed. "I'll put a plate in the oven for you," she said, and went back to the kitchen.

The ride from my home to St. Joe's takes about twenty minutes. I drove west on Indian School Road, squinting into the glare of the setting sun and wondering what could have gone wrong. Failures of this sort make up only about 2 percent of outcomes, but each time I confronted one, my earlier shortcomings rushed back to haunt me: that first year in med school when I'd scored my first "C" ever on an exam; failing the first part of the USMLE; and the awful day I'd been informed that I had gotten a "D" in my pediatrics rotation at University of Arizona med school. Anticipating the difficult conversation with my patient, I couldn't help but think ruefully of the reason I had failed that class.

That episode had been the closest I ever came to flirting with self-annihilation. The peds class involved fake interviews with another student pretending to be the patient, or in the case of certain fields, the family member or person representing the patient. The specialty area where the interview was critical was in peds, where we had to interview the "mother." The interviews were videotaped and evaluated later on.

The most sensitive part of the interview was giving the "mother" a diagnosis on her child. This had to do with the bedside manner that we were developing, which was a critical part of a med student's training; consider how you might deliver the news to a mother that her child had leukemia or some other terminal disease. I thought I was developing good communication skills in this respect, but it was the pediatrics class and the interview portion in particular that nearly did me in.

I had done the work in that area under Doctor Fulginiti, the chairman of pediatrics at U of A. I had held my own on the exams. I was feeling good about getting another rotation under my belt, and was looking forward to the party some of us were planning to celebrate the end of the clinical rotations. I think it was a Friday afternoon when I was called into Dr. Fulginiti's office. The assistant chairman was there with him. He didn't waste any time before he dropped the bomb.

"We called you in, Sonntag, to let you know you're getting a D in pediatrics," Dr. Fulginiti said. "It's your interview skills. We just don't feel that you can communicate well with the patient's mother."

I looked at my professor in dismay. Behind his dark eyeglass frames, his eyes invited no discussion; nor did his pursed lips. I don't remember if I objected or even what I said in response.

All thoughts of going to the party disappeared. How could I face my friends? I got back in my car and headed west on Speedway—the main drag through Tucson—not knowing or caring where I ended up. When I came to the city limits, I just kept driving into the desert, past the desert museum, into the mountains. The sun sank lower in the sky. My eyes kept veering from the road ahead to the edge on one side, where the land fell away into dry gullies and rocky washes. In some places, canyons and deep valleys yawned below. I knew I wouldn't do it, but I thought how easy it would be to let the wheels take me over the edge.

I thought about the thousands of dollars I owed on student loans. I thought back to the low grade I had gotten that first year. Maybe I should have realized then that I didn't have what it takes to become a doctor. I thought of quitting. I had no idea what I was going to do next. I had never thought about any other future outside of medicine.

Somehow I got myself home that night. I holed up for a few days. The news leaked out. The three other students who had done the pediatric rotation with me were only slightly less shocked than I had been.

All three protested the grade I had received. Nothing came of that, but it did help me to know that my fellow students backed me up. There was nothing else to do but keep going.

And keep going is what I had done. But from time to time I wondered about my communication skills. Was I lacking in compassion? Was I too curt? Was it my accent? For the sake of my ego, I had retaken that peds class and passed it with high scores—interview portion and all—but I still wondered how I came off to my patients.

One thing I knew now: Whatever position the nurse occupied on my "difficult patient" scale, I couldn't blame her. Something was wrong, even if it was a poor response to the pain medication. I would return and deal with the problem with patience.

It could be discouraging, though. It seemed I had been dealing with human frailty forever. There had my father's brain abscess and my mother's anxiety, weight loss and—at the start of my freshman year in high school—her broken hip. Those experiences had surely helped develop in me a necessary toughness, an ability to maintain focus in the face of other people's pain. They had also served to remind me that every case had a human drama behind it, and that my manner with the patients I saw could either assuage or exacerbate their pain. I did not want to be so removed that a kind of callousness set in.

Shaking away the ghosts of the past, I pulled into the hospital's parking garage. The screaming nurse was waiting. She had a story too. She was in pain. It was my job to find out what the problem was and to comfort her.

A moment later, however, the first verbal shell blew a hole through my good intentions. My patient's trumpeting invectives reached my ear from clear down the hall. As I grew closer and could make out the words, my own ears burned with embarrassment. The only other time I remembered hearing such crude profanities was during a stint I'd had loading trucks with the Teamsters the summer before starting med school. I think even those rough-and-tough blue-collar guys would have blushed to hear what was coming out of this woman's mouth.

"Thank God, you're here," a nurse said, hurrying out of her room. "She's been hollering till you'd think her lungs would collapse. Cussing you out, cussing out the nurses and staff. I've never seen such a show."

I walked in mid-holler, a bit fazed, to say the least, but still ready to minister to her need.

"You," she screamed, spittle spraying. "I don't know what you fucking did to me but I want you to get me the fuck out of this goddamned fucking place."

I moved closer to examine her and try to calm her down. She wouldn't have it. "Don't you fucking touch me," she screeched. "You sunuvabitching, goddamned, fucking prick! I want to be taken to my own hospital right now. NOW! NOW! These cock-sucking morons don't seem to understand that. And then I don't want to ever lay my fucking eyes on you again."

Given her high agitation—she looked like she was going to blow a gasket—the best thing to do was to comply with her demands to be transported to the hospital where she worked. I got on the phone and arranged immediate transfer with a doctor at Scottsdale Memorial. I never did find out what the problem was. And I never got to test my bedside manner that day. But by the time I returned home that night, I was too unsettled to care.

CHAPTER 20

THE CELEBRITY PARADE

If you modestly enjoy your fame you are not unworthy to rank with the holy.
JOHANN WOLFGANG VON GOETHE

JUST BEFORE CHRISTMAS in 1992, I was scooting past the secretaries' desk on the way back to my office when I stopped short. Something had changed. Next to my medical secretary, Kelly, sat an unfamiliar, plump, young woman with dark tawny hair. She looked up at me—homing in on the full surgical bonnet I was wearing—with frank, appraising eyes. "This is Cheri," my assistant said. "She's a stay-at-home mom, but she's going to help us out part time for a few months."

"Oh," I said. "Good to have you here. Where are you from?"

"Home," she said.

I snorted. "I see how it's going to be," I said. "A smart ass, eh?"

Cheri turned to Kelly, her eyes crinkled with amusement. "I like this guy," she said. "We are going to get along just fine."

That was the start of a nearly twenty-year relationship with Cheri Ebert, just one of the hard-working office staff who brought order to a constantly changing office. At the end of her temporary three-month gig—during which all eight of the associates shared her services—I arranged with administration to have her to work exclusively for me. Organized and efficient, Cheri made herself indispensable, and a year into it, I convinced her to go full-time.

I must admit Cheri drove me a little crazy at first; I think she saw me as a hard nut to crack, a task she was determined to accomplish with friendly banter: *Where you going this weekend, Boss? Got any plans this evening? How's the family?* In the beginning, I warded her off with a semblance of indignation: "None of your business." But her strong work ethic and obvious concern for the patients earned my trust, and as anyone who has relied on support staff knows, when you find a gem, you do your best to keep them.

Cheri—and my other long-time assistant, Debbie Nagelhout—not only handled the scheduling, travel arrangements, and other clerical tasks, they also dealt with patient-management nightmares. Take the queen's stay. It was Cheri who fielded the flood of incessant calls; concierged the finicky requests and demands (brand new clean linens, no Jewish doctors); arranged security passes for anyone going near the seventh floor; placated my other patients (one of whom complained to Cheri that if she, the patient, were the queen, she would get seen immediately); and tried to shield me from unnecessary, time-sucking distractions. For her part, Cheri also took to mothering the fellows.

For these and other herculean efforts, Cheri and Debbie, and other staff, received effusive expressions of thanks. By the time the millennium rolled around, they also enjoyed an even more appealing perk: unusual proximity to a range of intriguing patients, not the least of them some well known, hunky celebrities. It was one of these—a Heavy Metal lead singer I'll call Lars Frost—that unwittingly set off a red alert connected to a piece of lethal weaponry that appeared without warning on my desk one morning.

I had just arrived in the office that day when, opening the door, I immediately saw a large brown paper bag on my desk. I opened it and pulled out the bulky object within. When I realized what it was, I nearly burst a blood vessel. In my hand was the wooden stock of a rifle with the words "Kill the Bastards" stenciled on one side.

I scrambled out to the desk where Cheri sat. "Call Security right now" I said. "It's an emergency."

While we waited, I showed her the gun butt. Her face blanched. "Oh my God," she said. "Who would do such a thing, Boss?"

The security guy arrived and we ran through different possible scenarios. "Could it be a crazy patient?" he said. "Any residents or doctors holding a grudge against you?"

At nine o'clock the phone rang. Cheri picked up and a moment later handed me the receiver.

"Hey Chief, good morning" the voice said. "It's John Miller, the radiology tech? I talked to you the other day about getting me an autograph."

"What autograph?" I said. "What are you talking about?"

"Lars Frost," he said. "I asked you if you could have him sign something for me. I left it in a bag on your desk last night."

"That's what this thing is?" I said. "The rifle stock? That's what you want him to sign?"

"Yeah," he said. "That's it. It's already got the name of his first album on it. I'd really appreciate it."

"Yeah, yeah," I said, feeling my heart start pumping again. "I'll see what I can do." I hung up, shaking my head. "Emergency's over," I said. "It's a friggin' souvenir."

I hadn't even known who Lars Frost was the first time I saw him in the examining room, a fit, fair-haired Viking with cropped reddish hair, bright blue intelligent eyes and muscular arms so heavily tattooed it was hard to see the color of his skin. The name on the chart meant nothing to me until Cheri set me straight.

Later when he came in, I had Cheri bring the gun butt. By then she had endeared herself to my famous patient by locating a rare children's video for his two small children, "The Brave Little Toaster."

"I hope I don't get into trouble for this," I said, passing on the request for him to sign the gunstock.

Lars rubbed his hand over the wood and laughed. "No problem," he said. "Happy to oblige."

My famous patient's treatment continued for several weeks. Every time he flew over from San Francisco for his check-ups, Cheri displayed her usual over-the-top caring ethic. This included preventing any unwanted intrusions by putting a DO NOT ENTER sign on the door during his appointments. At the end of it all she was rewarded with a huge box of signed CDs, T-shirts, records and other memorabilia.

I certainly didn't begrudge her that. To tell you the truth, my patients' generous manifestations of thanks often embarrassed me, and when it was envelopes of cash—as with the Saudis at the end of the queen's stay—administration refused to facilitate the distribution of it. But I knew Cheri and the others were tickled by the gesture, and she, for one, deserved everything she got.

I had my own reward for a successful procedure that, given the value of my patient's vocal cords, had entailed a posterior approach rather than the usual anterior approach to his surgery. When Lars came for a last post-op visit, I ran through the usual simple tests.

"How's it going Lars?" I said. "Any progress with that arm?"

He raised his left arm, the one that had been unaffected before his surgery. Then scrunching up his face, he made a show of animating his limp right arm. He hadn't been able to raise it at all before the surgery, and it looked like there had been no improvement. But before I could let out a gutter sigh of disappointment, he held up the arm with the other one in a flapping wing motion. "Da da!" he said. "Looks like the old arm's coming back Doc."

That was reward enough for me.

The heavy metal singer was just one among a steady stream of celebrity and VIP patients who came to my door from the mid 90s on. Among those patients, I counted a number of nationally renown sports stars; a voluptuous Mexican soap-opera and daytime TV diva who fell off an elephant while filming a segment of MTV; a top Israeli military leader whose gaggle of body guards made me thankful that his C1-C2 fusion came out well; and a stream of wealthy Israelis and Hassidic Jews channeled through a rabbi in New York City.

One case that really got all of us hopped up was a rock megastar, notable for his New Jersey working class ballads, who needed minor treatment before going on stage one night during his tour stop in Phoenix. Relaxed, still ruggedly handsome, he was the nicest guy you could ever meet. Near the end of the consultation, he stood, pulled out two tickets from the pocket of his jeans vest, and said, "Hey, can you come to my concert tonight?"

"I'd love to," I said, taking the tickets. "I'll just have to tell my son that I'll miss his lacrosse game tonight."

"Your son plays lacrosse?" he said.

"Yeah, he plays lacrosse in high school."

"You can't miss a lacrosse game," he said. "My son plays lacrosse. I never miss a game when I'm in town."

I still had the two tickets in my hand but before I could think how to get out of the pickle I'd gotten myself in, he snatched them back. All I had time for was a last observation:

"You know," I said, "your back would be much better if you didn't have to do all that stuff you do on stage."

He laughed. "Totally. But that's part of me. I've got to do that," he said.

On his way out, he stopped at Cheri's desk. I heard her jib-jabbing with him, relating some story about her daughter's musical inclinations. A short time later she came sashaying into my office. "Guess where I'm going tonight?" she said with a triumphant smirk.

"Where?"

"I've got two free tickets to the hottest concert in town."

"What?" I said. "How?"

He told me he'd give me the tickets if I delivered his medical records backstage."

"No kidding, I said ruefully. "I want to go too, darn it, but I already told him I've got to go Stephen's lacrosse game. If I don't, he'll think I'm a bad dad."

So, unlike Cheri, the radiologists, and any other staff who helped out that day, I lost out on the concert. I did get my own back at Cheri down the line, though (at least that's how she and Debbie saw it) when I passed on a chance to consult with perhaps the most beloved leading man of the last forty years. I can't name him, but his facial crags and charmingly twisted smile have become legend.

I'd gotten a call from the movie star's doctor, and as I always do I first needed to see my availability. "Check my schedule," I said to Cheri. "Should we fly him out here?"

"Oh yeah, do it, Boss," she said.

As it turned out, it was nothing I couldn't handle over the phone. When I told her it was a no-go, she gave me the sour mug treatment, and for several weeks I heard a lot of grumbling about gypping her out of a chance to meet him.

Those were a few of the notables, but their stories were not the only colorful ones. Lots of ordinary folks made a splash, like the local "exotic dancer" who came in with a herniated cervical disc. I remember my fellow going in before me to take her history during the first appointment. "Chief," he said, fanning his face when he came out. "I think you're really going to like this one. She's an 'entertainer.'"

Cheri rolled her eyes. "Entertainer, my ass," she said. "She's a stripper, Boss. She was probably wrapping around the pole too hard."

However she defined her professional services, the patient was a very sexy blonde named Kiva or Kiya—something like that—with legs that went on for days. She certainly provoked a sudden and intense need among the residents for a visit to my office whenever she had an appointment. In the end, I was happy to oblige her desire for a posterior keyhole framenotomy in the back of the neck, rather than the usual anterior job. She did not want to mar her considerable assets.

Then there was a tall, lanky kid—fifteen or sixteen years old—who managed to break his giraffe-like neck on a roller coaster at Magic

Mountain, and the wife of another rock lead-singer who broke hers when hit by a giant wave while strolling on a California beach. You'd be amazed at the variety of ways people can break their necks.

Still, though all patients bring some level of stress, celebrity patients did add considerably more tension to the mix, requiring both extra protection of their privacy and "skin-to-skin" treatment. With them I not only performed the crucial part of the surgery, but also opened and closed. That was par for the course on my end, as were the signed portraits, gifts, free passes and other perks on theirs.

In that last respect, no one outdid the oil-rich Middle Easterners who came to my door or, as with the queen, summoned me to theirs.

Even before I had operated on the queen, a floodgate opened up from the Middle East allowing a steady flow of wealthy patients from that region to find their way to Barrow. Until 9/11 ended that flow, I was getting regular referrals from the Gulf countries. There was a haughty and demanding Saudi princess the year before the queen; the wife and sister of a high-born cardiologist from Egypt; the Crown Prince of Qatar, who gave me a glitzy watch to add to my collection (luxury watches make very popular gifts); and around the turn of the millennium another Saudi prince who flew Lynne, my younger son Stephen, and me first class to Paris. For two hour-long consultations with that client, I earned a hefty fee as well as unlimited limousine service complete with a guide and special passes to the Louvre and other tourist locales; two rooms at the spectacular Bristol hotel; and the absolutely superb food and amenities you might expect at such a place. It was Christmas time, I remember, and the pine trees along the Champs Élysées were strung up in white lights.

Then, in 1998, even before the queen died, I got a call from yet another Saudi prince who wanted to know if I could pop over to Cannes to examine him. I almost declined that request. After my experience with the queen, and given that BNI was by then flush with patients, I had decided not to go off on any more trips involving private patients. I was just explaining this over the phone when I noticed Cheri gesticulating wildly at me and mouthing a point that she obviously thought was important for me to consider.

"Remember Mr. Chalmers," she said, referring to another patient of mine, Herman Chalmers, who had made his wealth in construction. "He wants to give you those two tickets to a World Cup match. It's in Paris this year." Then she drove home the clincher. "It's Germany against the US. Why don't you combine the Cannes/Paris trip with your upcoming trip to Hanover?"

I have to admit that my passion for soccer clinched my decision when it came to accepting that case. It would be my second World Cup, after two matches at the 1994 games, and the first time I would get to see Germany play.

Just thinking about it produced the same excitement I had felt as a boy of ten listening to the final match of the 1954 World cup between West Germany and Hungary. That is something I will never forget, huddling around the radio with Gunther and his friends, listening to the static roar of 60,000 spectators at the Wankdorf Stadion in Bern. Germany had only been reinstated as a full FIFA member four years earlier, and though the postwar economic recovery was in full swing, the country was still coming to grips emotionally and psychologically with its Nazi legacy. Germany—the disgraced underdog—went on to win its first World Cup that year. In doing so, they moved a step closer toward taking their place again among the community of nations.

Given my passion for soccer, I accepted the prince's offer. I wanted to share my good fortune with my brother Gunther, but he was out of the country, so I called up my cousin Heinz Peter, who agreed to meet me in Hanover. There I took care of the several talks I had been invited to deliver at a neurosurgical conference. The morning following the conference, Heinz Peter and I boarded a private jet—we were the only passengers—and settled into two of the six leather captain seats in the roomy cabin. We enjoyed the full breakfast that had been laid out for us as we peered out the windows at the icy blue peaks of the Alps far below. When we arrived in Cannes, a limousine was waiting to take us to an elaborate villa high in the hills with panoramic sea views. The prince was expecting us. After the consultation, we had an elaborate lunch followed by another ride in the limousine down to the harbor, where the prince anchored his yacht.

It was a sleek, white whale of a vessel; Heinz Peter later learned it was then the fifth largest in the world. The visit is a blur now, but I remember a huge disco dance floor, an Olympic-sized pool, and a luxury hot tub that must have been 25 feet long. The prince also showed us a mammoth refrigerator stocked to the gills with beer. "But you're Muslim," I said to him. "We have many Western guests," he replied. The tour ended, the prince and I parted cordially. Heinz Peter and I were immediately taken to the airport and flown to Paris in the jet.

That night, my cousin and I saw the World Cup match between Germany and the United States. It had been nine years since the Berlin Wall came down. The excitement and patriotism exhibited by both sides were phenomenal. The USA was outplayed in that game, losing 2-0, but for me, it hardly mattered who won. I thrilled to the fervor that both sides exuded. In a sense, "my side" couldn't help but win, since both Germany and the USA claimed a piece of my heart.

The next morning, we were picked up again, taken to the jet, and flown back to Hanover.

The perks were heady indeed, as was the boost that working on all those VIPs gave to my ego. But there was a down side: the spike in anxiety such responsibility provoked, generated not only by the fearsome responsibility that pressed on me each time I walked into the operating room but also by the slew of other duties that required an almost superhuman self control, energy, and focus.

I had gotten my first inkling of the demands my profession had in store for me back in Boston working under Ben Stein. I had thought I knew what I was getting into during my residency, and I was used to working hard, but if someone had pulled me aside then and said, *Hey, you could walk away from this, you could be a restaurant manager and have a good life*, I think I would have thought hard about quitting. Even then, I was pushed to the limit, stressed emotionally, mentally, and physically. When, during that period, I came into to work one day to find that a neurology resident had committed suicide, I wasn't at all mystified about what had driven him to it.

Beyond the stress of the work itself, there were also other repercussions from a bad outcome. If the surgery or treatment didn't go the way I wanted, I not only had to worry about my reputation in the neurosurgical community; I also had to worry about some news reporter yo-yo sniffing around for a story to blow up with wild exaggerations. I could just imagine the headlines: "Valley Neurosurgeon Cripples Beloved Entertainer For Life"…"Botched Operation Puts Beloved Sports Figure Out of the Game Permanently." Other than exerting all the self-control and mental discipline I could muster, there was nothing to relieve that level of anxiety, except a kind of shutdown.

It had been a typical week: about ten surgeries scheduled; office visits with patients; rounds with the residents; journal articles and textbook chapters to complete or edit. I was at Saint Joe's. It was about nine-thirty in the morning. I was so used to operating on autopilot, performing one task after another without stopping to monitor myself, that now the existential moment at which I had arrived was a void before me. I felt like a manic drill that someone had pulled the plug on.

Around me the hive buzzed on. Bells dinged; wheels ruttered down the corridors; voices rose and fell like waves. Still in my scrubs, I walked to my office, grabbed my briefcase and keys, and scuttled out of the hospital before anyone could waylay me. I drove home, my overloaded head gone haywire. I couldn't remember if it was a tennis day for Lynne, or if she had a committee or school meeting. *Just let her be home*, I thought.

I tore through the living room and kitchen, then took the stairs on the double. Lynne was in our bedroom, pulling up the sheets on the bed. Sunlight streamed through the window. A calm blanket of quiet enveloped me.

Lynne turned when I entered, a quick alarm in the pitch of her voice. "Volker," she said. "What are you doing home? Is everything okay?"

I stood in the middle of the room, suddenly realizing how shallow my breath was. I gulped for air. I had no idea what I was going to say, but when words came, they were the only ones possible. "I can't do this anymore," I said, before the surprise had faded from Lynne's face.

"What do mean," she said. "What are you talking about?"

"I just cannot do this career anymore. I can't get in my car in the morning. I can't go down to Saint Joe's. I've just had it."

Lynne breathed deep, sat on the edge of the bed, and heard me out. Then, when I had stopped pacing around the room like a madman, she gave me that sensible look of hers (a look I knew all too well) and put voice to what I already knew. "You have patients waiting for you, Volker, that you can help."

That was the irrefutable truth and reality of my life. I rubbed my forehead. I looked at Lynne, the woman who had willingly taken on this responsibility with me. "Yes, of course, you're right," I said. I returned to my patients and finished out the day.

In the end, that's what sustained me: the knowledge that what I did everyday spelled the difference between health and illness, life and death for my patients. The perks were great, but the best perk wasn't what the wealthy bestowed; it was the satisfaction of healing people, saving people. It was also being in a position to pay back a deep debt to my adopted country, and to those who had helped me.

A few incidents illustrate this trade-off: Not long after my minor meltdown, a man stopped me in the aisle of a local food store. "You're Dr. Sonntag," he said, reaching for my hand. "You operated on my son. You know, you basically saved his life." A week or so later, I was in the hospital when I ran into a man at the hospital whose neck and back I had operated on. "Dr. Sonntag," he said, also taking my hands, "you changed my life."

That kind of feedback constitutes the best part of being a neurosurgeon. You walk on air when, after a successful surgery, you are able to pass through those doors to where the family is waiting and say: *Everything looks good; we got the tumor out; your loved one will hopefully be home in a day or two; I think everything's going to be okay.* No amount of money in the world can beat the satisfaction that comes from that.

Around the same time, a wiry, work-worn man came in for a consultation. It was Mr. Reynolds, the chicken farmer who had given me my first real job. I hadn't seen him in thirty-five years. We talked a little bit about the old days. "You made something of yourself," he said. "You were never afraid of hard work. I could see that in you from the start."

The consultation was short. It was not something for which my old boss needed surgery. I walked with him from the examination room back to the front desk and exit. "There's no charge for the consultation today," I said. "It's a small thing. I'd like to do that for you."

He looked at me, his eyes narrowing. I recognized the pride and self-reliance that had gained my respect years ago, mixed, I thought, with an edge of indignation. "Not necessary," he said curtly. "Absolutely not."

That was the last time I saw him. It was a fitting postscript to our relationship, a final reminder of the honor that belongs to a man who paid his way, and gave me an early lesson on how to pay mine.

CHAPTER 21

THE BLUE RIBBON

The price of success is hard work, dedication to the job at hand,
and the determination that whether we win or lose,
we have applied the best of ourselves to the task at hand.
VINCE LOMBARDI

I STOOD AT THE podium with a lump in my throat. It was 2002 and I was in Philadelphia to receive recognition as Honored Guest at the 52nd Annual Meeting of the Congress of Neurological Surgeons—the highest honor a neurosurgeon can achieve. Because it was an honor bestowed on me by my peers, it carried something of the prestige of winning an Academy Award, or rather a SAG Lifetime Achievement Award. If anything drove home to me how extremely fortunate I was to have found my calling and to have been able to contribute as much as I had, being so honored by the CNS was it. Presiding over the event was my third fellow—now president of the organization and good friend—Stephen Papadopoulos. Sitting in the crowd before me were my beautiful wife Lynne and our three children, now aged twelve to twenty-four.

That week of talks and meetings was the highest point in my career; I could barely contain my emotions when I thought of the road I had traveled—one that had taken me from a postwar refugee camp to the

pinnacle of my profession. I was especially thankful that I had recognized the critical importance of family to whatever success I had achieved.

Certain deprivations as a boy had done much to instill in me the commitment to marriage and family that saw Lynne and me through the long stretches away from each other.

With Lynne in Philadelphia. Being recognized as 2002 Honored Guest of the Congress of Neurological Surgeons was a high point of my career.

Many neurosurgeons' marriages, especially first ones, break under the strain. A cluster of memories rose in my memory, of experiences in stark contrast to one another.

One memory involved an evening during the spring track and field season my freshman year in high school. We had approximately nine meets that season, and I was fortunate to come in first or second in most of them. That led me to qualify for the state meet in the freshman mile. Three individuals were favored to win the event, a student named Becker from Camelback High, another from South High, and myself. As the last lap unfolded, I was in third place. Rounding the final curve, with approximately 100 yards to go, I had moved up to second. All I could think was that with all those practices, all that sweat, and all that energy I had put into it, I was going end up in second place. That fired me up. I gave it everything I had. I inched closer to Becker, who was in the lead. As the finish line closed in, I lurched forward, throwing myself about a foot in front of my opponent. Totally spent, I crumbled to the ground and had to be carried off the track like a fallen warrior.

I was proud of that blue ribbon. It didn't even bother me that I had to go to work at the chicken farm that evening, since that meant I could show it to Mr. Reynolds. My boss slapped me on the back, and when he called his two boys over to see it, my feelings of loyalty to him went up a notch.

When I returned home later that night, I was still pumped up about my victory. I found my parents in the dim living room watching television.

"I set a record today," I said, pulling out the ribbon to show them.

Neither of them looked at me. After a moment, my mother spoke: "Volker, take the garbage out," she said.

I left them and went to the cramped room I shared with my father—my mother suffered from anxiety and insomnia, so this had been the arrangement for several years. I pulled out a bottom drawer, buried my prize among my few other mementos, and headed to the kitchen to do my chore.

I don't think I blamed my parents even then. That's just the way things were. They managed to attend a few key events in my young life, but most of the time they were just surviving. And that's what they expected my brothers and me to do too.

Now looking back, I realize that it wasn't that blue ribbon that mattered anyway. It was the running. And though I had no inkling of it then, running now seems an apt metaphor for my long-term goal of becoming a doctor. In both endeavors, it would be the endurance that would get me there. In both endeavors, there would be plenty of small successes completely unrecognized but crucial in the long run. And in both endeavors, there would be bursts of focused energy punctuated by periods of waiting. The only difference is that, years in the future, when prepping for an operation, all nervousness would fall away, and a calm focus would come over me. Back in high school, waiting to compete was nerve-wracking.

I recall a soccer field under a full autumn moon thirty-five years later. My son Christopher was maybe ten years old. He looked like me at that age, with his compact frame, his close-cropped rug of curly hair, and the determined set of his features. We were on a practice field at a local school, surrounded by members of Chris's soccer team, which I coached. The days had gotten shorter, and a crispness in the air raised goose bumps on our skin. The sun had gone down in a flaming light show, and that impossible saucer of a moon rose huge and opaque over the majesty of Camelback Mountain, purple in the deepening dusk. I clapped my hands to call the boys' attention to the magnificent sight of the moon. We stood for a moment transfixed, oohing and aahing in unison before resuming the practice.

This memory is attached to the decade and a half I spent coaching Chris and Stephen in soccer and little league baseball. It was through those evening practices and Saturday games that I bonded with my sons and found the balance that sustained me during the long hours in the operating room and office.

Amid the surgeries and politics, the teaching and the legal hurdles, the publishing pressures and editorial work, my children were growing up. I strove to be a part of their lives, but it was a challenge. Now that I had achieved some success, I especially worried that my daughter Alissa, born just before we returned to Arizona, had gotten the short end of the stick when it came to fatherly attention. *Had I been involved enough?* I wondered. *Did she feel bad when I was away one third of the year? Did I miss something I should have been there for?*

Thankfully I underestimated my daughter, and recently she told me that it is my presence she remembers and not my absence. And that is what I remember, too: staying up late to bake cookies in the shape of Massachusetts with her for a fifth-grade school report; reviewing her essays over the breakfast table; sharing in her triumphs on the tennis court and bursting with pride at the bright, driven, beautiful young woman she was becoming in high school. I so easily could have missed out on those moments, and I would have if I hadn't realized early on that family was the fundamental structure upon which my happiness depended.

As we move through life, one thing that provides a clue to what we should value is lack of that very thing at other times in our own lives. In my case, although I understand the reasons for it, I remember with regret not having been able to share experiences like the blue ribbon with my parents. And now, it is not even the boy I was that I feel sorry for but the two of them. I find it sad that they could not set aside their troubles more often and fully enjoy what, for me, is the supreme happiness in life: being with my wife and children, whether at their school events, family vacations, or simply relaxing at home.

Still, my parents had loved each other and they stuck together. They had provided a good model of marital commitment, as had Lynne's parents. Long ago, when Lynne and I had announced our plans to simply live together and try things out, Lynne's father had immediately corrected our thinking on the matter.

That had been in April of 1974, when I was in my second year as a resident at Tufts New England Medical Center. I hadn't seen Lynne for over a year. We had exchanged letters after our whirlwind romance in Tucson, a blizzard of crisscrossing mail that petered out after a while to an unseasonable snowflake now and then. Our romance heated up again when she came out to Boston to spend a couple of days with me. Wanting to impress her, I took her to a restaurant with white linen tablecloths and ordered a bottle of Riesling. After that, the letters gathered force, like a strong wind.

A couple of months later, I flew back to Tucson to see her. Lynne picked me up at the airport. We stopped at the Green Dolphin, scene of my med school drinking days, where John the bartender said, "Hey, where ya been, haven't seen you around the last couple of days." It had been eight months. After that, I returned to Boston, Lynne carried on with her nursing studies, and a stillness descended over the landscape of our romance.

That April, I came back to Phoenix, not to see Lynne but to visit my father for what would be the last time. The lung cancer he had been diagnosed with had metastasized and moved into his brain. While I was in Phoenix, however, my good friend Sam Shoen called to invite me to a party on Easter eve. That sounded pleasant. I didn't think much about it but was looking forward to seeing my friends from med school.

It was like holding a match to a tumbleweed when I saw Lynne across the room. Once we began to talk, everything around us seemed to recede, and all the boxed-up words of the last year came out at once. We were deep in conversation until the early hours of Easter Sunday. Before we were through, after weighing all the pros and cons, we decided that Lynne would come back to Boston with me to try things out for six months. We also agreed that we would see each other the next evening, after the Easter celebrations with our families.

When I rang the doorbell at Lynne's house Sunday night, I was expecting to see her beautiful face. It was her father, though, who answered the door. I felt a stone hit the pit of my stomach.

"Hello there, Volker," he said. "Would you mind stepping out to the patio with me?"

"Sure," I said, and followed him out to the back of the house, where we settled into the patio chairs.

"So, I hear that Lynne is thinking of going back with you to Boston. I guess the plan is that she'll stay there with you for six months or so, get a feel for things."

"Yes," I said, "that's certainly what we discussed."

He raised his eyebrows before continuing, and his eyes held mine.

"What are your intentions?" he asked.

April is a cool month, but suddenly it felt like July.

"Um, well, if things work out after six months, we might get married."

I thought this was a reasonable statement, something that would appease a father.

He waited a moment. Nodded his head.

"Well," he finally said, "if you are thinking of getting married, then why don't you marry her now?"

There it was. There was only one way to answer that question. I went for it.

"We can certainly do that," I said.

With perfect timing, Lynne appeared at the door and came out to the patio.

"I think I just agreed with your dad that we should be married," I said.

"We have to talk," was all she said.

We drove off in Lynne's VW and ended up in the parking lot of a Circle K convenience store. We weighed all the factors. My father was still alive. It would make him happy to see me settled with a good "Sonntag woman." Both of our families were in Phoenix. And, a significant factor indeed, her father had asked us to do it. How could we go back now and tell him we were just going to live together? It was settled. We went back to Lynne's house and made the announcement.

We were married the following Wednesday and back in Boston on Thursday evening, where Lynne sat in my ship's galley of a kitchen looking dubiously at walls the color of chewed bubble gum and a refrigerator empty but for a single can of beer. On Friday I was back on call at the VA and Lynne was on her own with no friends, no place to go, no one to even call.

That was forty-two years ago. Without Lynne's father's veiled mandate that we do it right—that we get married—we might have started out without the commitment that is essential to marriage, and that saw us through the hard times ahead of us. It's true that commitment takes different forms with different couples. But if there is one point I drive home to the residents I teach as they struggle to find balance between a demanding profession and their personal lives, it's this: Taking the time to be with family—however they conceive of it—will be a source of either peace or regret in their hearts in years to come.

The Congress of Neurological Surgeons took place at the Philadelphia Museum of Art, a key location for the movie Rocky. Being the honored guest at the CNS that day, I felt just like Sylvester Stallone's character in the famous scene where Rocky runs up those seventy-two stone steps. At the end of the weeklong event, I ran up those steps myself, the Rocky theme song ringing in my head. At the top, I turned to take in the magnificent cityscape and, like a million tourists before me, raised my fists to the heavens and leaped into the air. It was a great feeling.

CHAPTER 22

MALPRACTICE

Malpractice: a dereliction of professional duty or a failure to exercise an ordinary degree of professional skill or learning by one (such as a physician) rendering professional services which results in injury, loss, or damage.
MERRIAM WEBSTER DICTIONARY

If people understood that doctors weren't divine, perhaps the odor of malpractice might diminish.
RICHARD SELZER

I NEARLY MADE IT through the entire wedding and reception without seeing the woman, though she was (and is) a close relative of the bride and groom, as Lynne and I are. The ceremony itself passed without a glimpse of her, as did the meal at the bride's home. She ate with her kin; I ate with mine; and the rest of the time Lynne and the children circled the wagons around me. They knew how painful it was for me to be in such near proximity. Only at the end did I see her, when she blew into the room where the bride was showing Lynne and me the wedding gifts. It was only a split second—I saw her out of the corner of my eye—but the sight of her triggered a visceral and immediate need to hit the road. *Friggin' beautiful,* I thought, for the tenth time, *she is now officially my in-law.*

Let's call her Liz. The blond, attractive wife of a man with whom Lynne had gone to both grade school and high school, she had come to me a year or so earlier—in 2004—for a routine cervical fusion. I gave her the choice of having a plate to stabilize the spine or wearing a collar for some days after the operation. Like the vast majority of my patients—I plated nearly everyone in the last decade of practice—she chose the plate.

Liz came through the surgery fine—the notes I later referred to corroborated this. When my fellow and I saw her in recovery, she took sips of water and spoke to us with no apparent difficulty. There was no sign of a complication of any sort. I relayed the good news to her husband, parents, and other family in the waiting room and went home thinking everything was well.

Between that post-op visit and my rounds the next morning, however, something changed. Shortly after arriving at the hospital, I saw Liz's father striding toward me in the hall. "What have you done to my daughter?" he said.

"What do you mean?" I said. "Has there been a change? When I left her in recovery yesterday, she was fine."

"Well she's not fine now," he said. "She can't swallow or talk this morning."

"It's normal for some swelling to occur at the incision," I said. "The esophagus can be affected too, which will cause discomfort, but that should diminish after a couple of weeks. Any serious problem is usually evident when the patient is extubated. We saw nothing to indicate a problem."

He looked doubtful. "I don't think you understand," he said. "She can't swallow or talk at all."

At that moment, Liz's husband marched up. He and his father-in-law could have rehearsed their spiel together. "What have you done to my wife?" he said, jutting a rather impressive handlebar moustache at me.

I instinctively backed away a step—the guy was about 6'4" and a real bruiser.

I repeated what I had told the father and then hurried with them to Liz's room. Her mother was with her. Liz looked worried, but not overly agitated. The residents had already ordered X-rays earlier that morning when Liz first complained of the complications. The films showed nothing that would raise concerns.

"The X-rays look fine," I said. "It's most likely delayed recurrent laryngeal nerve palsy. That's usually a temporary loss of movement in one of the vocal cords. As we discussed before the surgery, it's a well-known complication of the operation, though it's rare, and it usually gets better."

The next day Liz showed no improvement. We called in an ear, nose, and throat doctor to take a look at her. We ordered an esophogram to check the swallowing. The scan was not abnormal.

"The difficulty swallowing almost always decreases with time," I explained to Liz and her husband. "If that and the hoarseness don't get better after a month, the ENT doctor can inject the vocal cord with gel foam or other material to get the voice back. But it usually comes back on its own."

I felt confident the problem would go away. In my career I had had maybe six such cases, and in each one the patient's ability to speak had returned within weeks. In one case, we'd had to feed the patient with a tube for a number of days until the esophagus recovered, but recover it did. The ENT doctor concurred with my conclusions. The couple seemed satisfied. Liz was discharged on the third or fourth day following surgery, after receiving a speech therapy consultation to inform her on what she could eat and drink until her esophagus recovered.

Two weeks later I saw Liz for a follow-up. Through her husband, she claimed that her condition had still not improved. I told them to give it another couple of weeks. That was the last I heard from them for the next year.

Mistakes happen all the time in surgery, but when it comes to a patient's subjective experience of pain or dysfunction, barring direct physical evidence of a problem, it's very difficult to assign cause. The success rate of such surgeries as I had done on Liz, as well as the advances made in the design of surgical devices, both promised a positive outcome.

By the early 2000s, cervical decompression and fusion had been fine-tuned to a precise science: from the small transverse incision on the anterior neck allowing direct access to the damaged vertebrae and ruptured disc; to the use of two small Caspar pins on the two vertebrae flanking the herniated disc to pull the collapsed vertebrae apart; to the microcurettes and fine burrs that, respectively, allowed removal of the herniated disc and precise shaving down of the bone spurs—both in the front and back of the vertebrae and within the disc space itself; not to mention the evolution of increasingly sophisticated anterior cervical plates.

The plate had come a long way since I had operated on the queen in 1994. The Synthes titanium plate I used on her was fairly simple compared to the designs that emerged just a few years later. Even as the queen was recovering, Steve Papadopoulos, another neurosurgeon named Reg Haid, and I had started working with engineers at a surgical device manufacturer called Codman—later acquired by Johnson and Johnson—on improvements to the device. Later my two fellow neurosurgeons and I joined Danek (now Medtronic), and after two more years, in 1998, the first version of the Atlantis Plate came on the market.

The breakthrough on that first version was the development of a mechanism that would lock in place both fixed and variable bone screws; that is, screws that could be driven into the bone at different angles depending on the pathology. It took a whole year just to figure out that simple element, but what we devised, after much trial and error, was a locking mechanism in the form of a simple gold washer under the central lock screws. First, two bone screws were inserted per each vertebral level that needed to be fused so that the edges of the screw heads slipped under the washers. As the central lock screws were screwed in and tightened, the washers pinned the screw heads down, locking them in place. The screws could not back out or slip, and, especially in the case of two convergent screws whose points were angling towards each other, you'd have had to remove the bone in between to get the screws out. If I had had that plate on the queen, I wouldn't have had so much trouble with the outcome.

Still, on that plate, with openings for only the screw holes, surgeons complained that their view of the bone graft was obstructed. That led to a redesign that provided "windows" on the bone underneath and which became the second version of the plate—the Atlantis Vision.

There was one more problem to tackle, however, and that was the compatibility of the screw heads, which were different depending on whether they were the bone screws or the central lock screws. With the first two versions, the surgeon drove in the bone screws with one specialized screwdriver, and switched to another to turn in the central locking screws. The third version, the Atlantis Elite, standardized the screw heads, allowing the surgeon to put the bone screws in and, with the same screwdriver, twist in the central locking screws. A last remarkably innovative improvement was added to this, when a new quarter-turn washer was developed, eliminating the need altogether for the central locking screws.

I used the Atlantis Vision for Liz's cervical fusion. After not hearing from her for months, I assumed she had fully recovered. That was not the case.

It happened just like in the movies. I was in my scrubs and bonnet, hurrying down the corridor after a surgery. I didn't notice the man until he appeared abruptly at my side.

"Are you Doctor Volker Sonntag?" he said.

"Yes," I said.

"Here you go, Pal," he said, slapping a fat business envelope in my hand. "Consider yourself served."

I felt like I'd been Tasered. Given the nature of the delivery, I had little doubt of the contents of the letter, though with about 12,000 surgeries under my belt by that time—and five times as many consultations—I had never experienced such a jarring event in over thirty years of practice. I beelined it to my office, closed the door and grimly opened the letter.

If you've never been served papers in a lawsuit, consider yourself blessed. I scanned the pages quickly, recoiling as select words and phrases jumped out at me: maliciously disregarded; exhibited gross negligence; resulting in permanent injury; affecting the patient's social and family life; and so on.

I felt weighted in place, leaden with distaste for the tortuous process I knew lay ahead. Then I called my lawyer.

Thomas York's office boasted the signs of success that engender confidence in a client: a massive executive desk with double towers; a wall of shelves bearing handsomely bound legal tomes; tufted leather swivel chairs; tasteful art in expensive frames; and, in a separate room, a long conference table—all of it gleaming in hardwood solids and cherry veneers. York himself with his trim physique, tanned face, and thick head of prematurely silvered hair, seemed yet another element of his perfectly appointed environment. The malpractice defense trade had clearly been good to him.

On this occasion the piles of papers were gone, the notes from the endless meetings, the copy of the deposition, and the reports from Liz's lawyer all filed away. The lawyers for both sides had met numerous times. York had consulted with my insurance company. Figures had been bandied about.

"I think it's a good decision," York said. "It's in your best interest to avoid a jury trial. All she'd have to do is convince a jury that she'll be like this for the rest of her life, and then the sky's the limit."

"And the insurance company? They'll agree to the compensation?"

"Yes, we've spoken several times. They also feel it's better to settle. Again, if it goes to trial, who knows what a jury will do? I still think its 50-50 either way, but her lawyer will wring what pathos he can out of it. She can't talk, can't swallow. He'll have her parade her children up and down. The husband will testify that his love life is down the tubes. It's all conjecture of course, but I think we can expect something like this scenario. Her lawyers are looking at the same possibilities here. Asking themselves how the jury would decide."

"Wait a minute," I said. "What about the reports from all those specialists she went to? There must be a dozen listed in her deposition. What about the neurologist she saw? And the psychiatrist? Now why would she see a psychiatrist?"

"I think we should leave that alone," York said, rolling a pen between his fingers.

Bile rose in my throat. I felt like a trapped animal. "I want to be done with this thing," I said. "It rankles the socks off me, but if you think we should settle, I have to go with your advice. If that's the best outcome we can expect, then let's move in that direction."

So, given the family relationships at stake, I decided to avoid a long-drawn-out legal battle. I settled—for a sum I agreed not to divulge. It gnawed at my gut, though. Once you settle, even for the right reasons, the implication of guilt sticks to you like a pant leg to a static-charged sock. You can't shake it off. For months you lie awake at night, replaying the charges thrown at you until you want to jump out a window. You can get hostile in your thoughts, not only about the other party's lawyer but also about your own. I will never understand how lawyers buddy up afterwards. The prosecutor treats you as if you have personally killed your own mother and then rambles over to your lawyer. They slap each other on the back, and head out for a drink.

Lynne and I managed to stay clear of Liz in the decade following the settlement, but Lynne did see her again recently at another family event that I did not attend. "She looks remarkably healthy," Lynne told me later. "And she was speaking just fine."

That suit was a blow, and not only because of the personal connection. I was still mired in legal proceedings from another malpractice suit brought two years earlier by a woman whose husband had come in for removal of a tumor in his cervical spine.

Gary and Nicole Martindale first flew out from Tulsa, Oklahoma to see me in 2002. I remember a nice looking couple in their early forties. The CT and MRI images that Gary brought with him, along with the medical imaging report, identified a tumor inside the dura of Gary's cervical spinal column.

He sat on the examining table, a man in his prime, apparently healthy. "So it is a tumor," he said.

"Yes," I said, "it's a tumor. Hopefully it's benign, but there are always unknowns about the type. I'll have to get in there to see what's going on with it. In your case, there is no prior tissue diagnosis and you are medically stable, so surgery for biopsy—and removal—is the first treatment option."

"What if it's . . . not benign?" his wife said, obviously hesitant to say the word "malignant." Her hands were locked in a tight grasp and her eyes razored my face as if she could eke out of me what she wanted to hear.

"Let's do the surgery first," I said. "We'll do an excisional biopsy, removing the entire mass and taking care that no tumor cells remain. If anything shows up on the pathology report, we'll do further imaging to locate possible metastases."

A warm glow emanated from the soft light reflecting off wood shelves, framed photographs, and fine mementos from Robert's many travels. It was late in the afternoon after a typically hectic day. We both wanted to get home, but I had to get this one detail out of the way first.

"The MRI scan shows a large intra-dural, extra medullary tumor at the C2-C4 level," I said. "The axial view shows considerable compression of the spinal cord. I want to get him in the OR for a surgical resection right away."

"Let's see it," Robert said.

I handed him the film and he clipped it up on the view box. "The tumor's right on that vertebral artery," he said, referring to one of the two main arteries that thread down either side of the cervical spinal column. "That might give you some trouble."

"Right," I said. "That's where you come in."

Robert and I had worked together over twenty years and there was no one I trusted more with a vascular abnormality or vascular involvement of some sort. His brown hair had weathered, like shingles on a barn, and the fine lines that had always radiated from the corners of his eyes and from his cheeks to his chin had deepened. He was still lean, still a commanding presence.

"What's the plan?" he said.

"Surgical resection with a posterior approach and laminectomy. The tumor looks contained. I don't think we'll need to do a facetectomy for lateral access. But you can help me dissect it out of there in case the blood vessel gives us any trouble."

"Ok," he said. "Get it scheduled."

The preliminary steps went as usual: the incision on the back of the neck; the placement of the retractors; the separation of the soft tissue from the bone; and the laminectomy at C3 and C4. The dura exposed, under high magnification I carefully opened the semitransparent membrane. With the first prick of the microinstrument, a small stream of cerebrospinal fluid gurgled out. Robert suctioned it for a minute or two, while I continued to pick and tear at the delicate dura until the spinal cord was well exposed—fleshy, white, with fine tributaries of veins, like a length of raw shellfish.

The tumor, milky and opaque, was readily visible, pushing the spinal cord posteriorly on to the left side of the spinal column. The mass quivered in clear fluid, like an oyster in its shell.

"There it is," I said, increasing the magnification with the foot pedal.

"Considerable pressure against the spinal cord," Robert observed. "But it's nicely encapsulated."

I made a mincing movement with a bipolar coagulator along my line of entry, and with microscissors carefully opened the cauterized capsule of the tumor. Then, with Robert's help, I began to dissect the tumor off the wall of the spinal cord, gently prodding it out of its capsule with a micro Penfield 4, cutting and coagulating, sucking at it with the ultrasonic aspirator to get it out whole. First, though, I snipped a bit of tissue, put it in a small bottle, and gave it to the rotating nurse, who ran it over to the pathologist for a frozen section.

About five minutes later, as I was still prying the tumor out, the phone in the OR rang. The rotating nurse answered. "It came back as a spindle cell," she said.

I felt a collective sigh of satisfaction. "That's consistent with what we're seeing here," I said to Robert. "It has the characteristics of a spindle cell or a schwannoma."

He nodded. "It all fits."

It was very good news. A tumor that is aggressively eating into the spinal cord would be more consistent with a malignant tumor. This one, cleanly contained and separated from the spinal cord, was fairly easy to remove and thus consistent with a spindle cell tumor. And usually when you talk spindle cell, you're thinking schwannoma—a nerve sheath tumor made up of spindle cells—which is usually benign.

After removing the tumor, we closed the patient up. He went home after a couple of days. Everything was normal. Uneventful.

After several weeks, Gary Martindale's tumor returned. We took him into surgery again. Again, the quick report came back as a spindle cell tumor. Now, the quick report is necessary so that the family can be informed quickly, and it also influences the surgeon's decision on how aggressive to be with the surgery. We were suspicious about the recurrence, but we took extra care to get the new tumor out completely. Again, the patient returned to Tulsa.

A month later I returned from a weeklong, out-of-town meeting to find the usual pile of mail waiting on my desk. Thumbing through it, I came to an envelope from the pathology lab that sends the permanent reports. I did a mental survey of the most recent patients I had seen, since the permanent report usually comes back two or three days after a surgery. I was surprised when I saw Gary Martindale referenced. My surprise immediately turned to dismay. The tumor had not been a benign spindle cell tumor but an extremely aggressive tumor called a synovial sarcoma.

To this day I have no idea why I did not see the report until four weeks after the surgery. I don't know if it got lost in the mail. I don't know if staff misplaced it. I don't know if the lab was late sending it out. I only know that the first time I saw it was that afternoon in my office. During the intervening weeks, having the quick read on Gary's tumor and not hearing anything from the pathologist, we had still thought it was benign.

I immediately picked up the phone and called the Martindales. Gary answered.

"It's Dr. Sonntag." I said. "I'm sorry to say that your tumor was not benign as we had thought. I just got the report myself and the pathologist thinks it a synovial sarcoma, which is actually a very aggressive type of tumor."

The brief silence pounded in my ear. "Thank you for letting us know," he said then. "But why didn't you tell us earlier?"

The weight of his question pressed on me. "Usually with such a diagnosis the pathologist would call us. But I only now found out myself through the report."

"What about the discrepancy in the reports?"

"The permanent report is the one you should go with," I said.

That was the last time I spoke to either Gary or his wife.

A synovial sarcoma is a very bad actor. It is more common in males than in females and, tragically, most commonly afflicts the young. The peak of incidence is typically before the thirtieth birthday. Survival on a bell curve is only one to two years. Having to inform a patient of such a condition is to ring the first toll of a death knell.

Robert was just as shocked as I was. We mentally replayed the events. If there had been any sign that the tumor was malignant, we would have unquestionably followed up with an oncologist consult and work-up for metastatic spread. We knew, of course, that if you leave behind a couple of cells of a benign tumor, they will most likely die, but if you leave just one cell behind from a malignant tumor, it will multiply. We figured that's what had happened with Gary.

I didn't know what course the Martindales took until months later when they brought a malpractice suit against me. Only then, in discussions with my lawyer, Thomas York, did I learn that after his second surgery, Gary had gone to a prominent cancer center. There he underwent aggressive treatment, including chemotherapy and radiotherapy. He also apparently had an idea planted in his head that gave him and his wife not only unwarranted hope but also grounds to sue me.

"The doctors at the cancer center told him they could have dealt with the cancer successfully if he had come in for treatment earlier," York told me in one of our endless meetings in his office.

"That's not necessarily true," I said, leaning back in frustration from the conference table. I glanced at the nurse two chairs down, who was present to help York phrase the questions appropriately. I uneasily observed York's associate, another fixture at these meetings, perfunctorily taking notes; I hoped he was notating everything I said accurately.

"This is a very, very aggressive cancer," I said. "No matter what you do, it's especially malignant. Just look at the numbers on it. There's a very small survival rate."

"That may be so," York said. "But their argument is still compelling. They are claiming that you did not recognize that that the tumor was malignant, that you should have done so, and that you were negligent in not bringing in an oncologist. That's what the jury will hear."

"What's going on with the hospital and the pathologist?" I said. "They are named in the suit too."

"Oh." York shrugged sympathetically. "Their lawyers dropped the hospital. And the pathologist was employed by the hospital, so they dropped him as well. I'm afraid they've singled you out. The way your group works, you're an independent operator, so they can do that."

I had the sense of treading a swelling sea while a ship receded into the distance.

"I see," I said.

It shouldn't have been such a bitter surprise to me that I had been targeted as the sole culprit. From the long day I suffered through the deposition, the angle the Martindales's lawyers were shooting for was clear. They nailed me good on their underlying proposition: I was the captain of the ship, and the captain of the ship has to know everything that goes on. *Why did you get the pathology report so late? Was there a problem in your office? Even if there was a problem in your office, you're in charge of that. Don't you think it was important to know the permanent report on the frozen section? Who was the patient referred to, you or Dr. Spetzler or the pathologist? He was referred to you. You're the captain of the ship. You're the one who sees to it that everything runs well.* And so on…

A couple of years into it, when I thought I'd never sleep through the night again, York called me with even worse news.

"There's been a development with the suit," he said.

"Okay. What's going on?"

He cleared his throat. "Martindale died."

"I'm truly sorry to hear that," I said.

"Well, it changes the case."

"How so?"

"It's not malpractice anymore. His wife is now suing you for wrongful death."

The punch was like the kickback from a high-powered rifle. "What does that mean?"

"Well, it's more damaging," York said, the implication of his words dangling like a fish on a line. "I'll have a clearer picture after meeting with her lawyers again."

The case dragged on interminably. Three or four years of depositions, hearings, and negotiations followed. Liz's suit came and went. As with that case, York felt I had a 50-50 chance. He thought one revelation in particular might sway a jury in our favor: the couple had successfully sued another doctor for malpractice in a case involving a disabled child. Still, York only had to voice the worst case scenario to make me despair of prevailing.

"You want the jury to see the widowed wife?" he said. "And on the opposite side the surgeon who 'killed her husband,' a man at the height of his earning power, a man with a wife and two kids to support, a man on the brink of being promoted when the tumor was discovered? Then you've got those young children there too. They'll argue that his lost earnings amount to millions. It's going to be very tough to win. Juries may try to decide rationally, but hell, the emotions are bound to be high on this one."

York was certainly right about that last statement. Emotions did run high. And not just for the plaintiff. I felt sick over the whole thing. I replayed the surgery endlessly in my mind. Images of the young couple rose and faded. York's words echoed in my head in the dead of night.

I had been unable to prevent the death of a patient.

In the end, I lost my stomach for it. I wanted it to be over. The resolution came behind closed doors, my lawyer, their lawyer, an arbiter and my insurance company hashing it out. I don't know if Mrs. Martindale was present, but I was not.

I was in my office when York called to tell me the final details.

"I'm glad it's settled," I said. "Outrageous sum of money but I'm glad it's done."

And that was it: a sudden, strange clearing, like a storm moving out. Relief lapped up in small waves the rest of the day. I told Lynne the outcome when I got home that evening. I slept like a dead man through the night.

So, what is there to say about malpractice in the United States? Unless you are a doctor, or married to one, you probably have little idea of the money involved both in being sued and in maintaining insurance against getting sued. What you pay out for protection against that eventuality is determined by where you've ended up hanging your shingle. Malpractice insurance in Arizona currently costs about $135,000 a year. Before I retired a few years back, I considered myself lucky to get such a low rate. It would have cost me twice that had I built my practice in New York or Philadelphia or some other major city in the East.

Having said that, Arizona is a tough state to practice medicine in. Trial lawyers are powerful here, and unlike California, for example, there is no ceiling on what can be awarded for pain and suffering. Of the two awards a judge or jury decides on—one for the actual damage done by the alleged malpractice and a separate award for pain and suffering—that second one is the killer. Lawyers will always go after it because there is simply no way to quantify pain and suffering. When there is no limit, why not ask for five or six million?

It didn't help that out of the top ten medical specialties most often sued for malpractice, neurosurgery comes out number one. According to a study in the New England Journal of Medicine from 2011, nearly one in five neurosurgeons risks malpractice, with 19 percent facing a claim each year. Even more startling, it was estimated that by the age of 65 years, a whopping 99 percent of physicians in high-risk specialties had faced a malpractice claim. That's a pretty big chunk of the 17,000 medical malpractice suits Americans file a year. Who was I to beat those odds?

It is not just monetary considerations that come into play, though. Except for the patient and his family, no one feels as terrible about a bad outcome as the doctor in charge. After all, the ultimate purpose of a doctor is to take care of the patient, to heal that individual, restore function, eliminate pain, and to do no harm. So, I don't think most people understand how personally we take failure. It affects our private lives. It affects every thought we have. That may not be of any comfort to a patient who feels he has truly been injured at the hands of an incompetent doctor, but should be considered when you are dealing with questions of malpractice. The body is an extremely complicated organism. When something goes wrong, the issue of fault is not always clear, and it is rarely simple. Litigation, and the financial incentives that enter into the equation for both plaintiffs and lawyers, only adds to the complexity.

And then there are the moral ambiguities. Like the lawyer from Kansas who came to me for surgery on two ruptured discs in his lower cervical spine. The surgery went well. The patient did fine and went home.

It wasn't until after the operation that I found out he made his living representing victims in big personal injury cases; one involved people injured in the collapse of a bridge at a large hotel. He made a fortune representing the families of the people who had died. That would have been okay, I guess, but his real bread and butter was bringing malpractice cases against doctors.

Some months after his surgery, at Christmas, a package of Omaha steaks arrived, complements of the lawyer. Every year since, he has sent me a box of Omaha steaks for the holidays. Now, would I have taken the case had I known about the malpractice suits? It's hard to say. And should I still be accepting those steaks?

Fortunately I was sued only those two times, though I was named in a couple of other malpractice cases that were eventually dropped before the deposition phase. And I have been involved as a member of the professional conduct committee of AANS in one or two unsavory proceedings having to do with lawsuits or charges brought by one neurosurgeon against another of inappropriate expert testimony. Ethical issues of this nature can get thorny and keep you awake at night, almost as much as cases directly involving your own practice. You never forget you're dealing with someone's professional reputation.

There was a postscript to the Martindale case. Some months after the settlement, I got a call from Thomas York. "I got a rather unusual request regarding you," he said.

"Oh? What is it?"

"It's from Mrs. Martindale. She called me up. She wants to meet with you."

"To meet with me? What on earth for?"

"I don't know. It's highly unusual," he said.

"Well that's a kicker," I said. "What do you think? I'll have to take your advice on this."

"Say no," he said.

And so I did.

CHAPTER 23

DRESDEN

*A people's relationship to their heritage is the same
as the relationship of a child to its mother.*
JOHN HENRIK CLARKE

IN 2004, I traveled to Dresden as part of a combined convention of the American Academy of Neurological Surgeons and its German counterpart. As president of the American organization, I was expected to give a keynote speech on a topic of my choosing, one that concerned a particular passion outside of medicine, saving the whales, for example, or protecting old-growth forests.

After casting around for a topic I could get my teeth into, I decided to call on my own experience as a naturalized American born in west Prussia, in the eastern part of Germany, in the last days of World War II. I called the presentation "A Personal Reflection of the Cold War."

It was a PowerPoint slideshow, chronicling my experiences as a member of a German family caught up in the chaotic aftermath of the War. As images from that awful time shuffled across the screen, I described how my mother had fled East Germany with my brother and me in tow; how we languished for four grey years in a refugee camp; how my parents finally scratched out a toehold in the postwar recovery, and then got knocked from that hold when an abscess formed in a pocket of my father's brain; and how, finally, we ended up among other destitute passengers on one of the last boatloads of postwar immigrants to the United States in 1957. By the time I ended with a summary on how the developments in Germany since the end of World War II, especially those of the Cold War, had influenced the course of my life and that of all Germans, the air in the room felt like a single, collective breath being held.

I am an American citizen now, but I know the German people. We have a reputation for being about as warm and expressive as cold fish. As I've recalled, my own mother and I never said we loved each other until the night before she died. Because I know this about the people among whom I was born, what happened next astounded me.

When I had finished speaking, a German member of the audience stood up, then another, and another. Soon the entire audience of three hundred was on their feet. Their clapping resounded through the hall like a hard rain on a tin roof. When the applause died down, several guests made their way to the podium.

"Very fitting, Dr. Sonntag," one said, his voice breaking. "I have not thought about those days in a long time."

"How your story resonates with my own," said another. "Very few people have ever talked about what happened to us after the war."

I was immeasurably moved, but it wasn't until one of the guests—a department head from a large hospital—took my hand that I felt my own throat closing up. Blinking repeatedly through eyes swimming in tears, he struggled to speak. *"Ich danke Ihnen,"* he said at last. *"Ich bin Ihnen sehr dankbar . . .* I thank you. I'm very grateful to you."

It seemed that what had happened to my family was a piece of a larger story that many Germans of my generation have been unable to tell, or even to explore for themselves. After all, our stories stemmed from our family histories, and who wanted to hear about the hardships German people faced after the war? Who wanted to hear how the generation of Germans who had brought the Nazis to power overcame adversity, did good work, loved and sacrificed for their children?

Even more: Perhaps my German colleagues wondered, as I did, at the path their lives had taken. Perhaps that day in Dresden, as part of an elite group of professionals, they were asking themselves the same questions that rose up in my mind: How had it happened that, six decades after that merciless firestorm devoured Dresden, I found myself sitting in that castle, itself glorious proof of Germany's renaissance, as a member of a profession that, in the United States, only a sliver of the population successfully enters?

I have returned to Germany a dozen times over the years, but it was the summer between my sophomore and junior years at ASU that I visited the first time since immigrating at the age of twelve. It was 1965, when you could do Europe on the cheap. I wandered around London and Paris for a week before heading east to visit my mother's mother, Oma, in Württemberg. My grandparents had been left behind on the Communist side after the war, and the East Germans had only allowed my grandmother to leave

Güstrow for the West after my grandfather's death. I saw my Tante Ev and Ev's son, Jörg Albrecht, and talked about family history; Jörg and I shared a connection to my mother's brother, Karl, the doomed, young Luftwaffe pilot who had gone down in flames, each of us having one of his names.

I did a short detour down to Italy, where I marveled at the great medieval *Duomo de Milano*. And when I stood before Michelangelo's *David* in The Gallery of the *Accademia di Belle Arti* in Florence, and saw for myself how the artist had captured the perfection of the human form, something deep and mysterious stirred in me.

At last, the time came to return to my past. I left my companions in Rome and bought a cheap ticket for the train to Stuttgart, then hitched a ride to Bad Hersfeld. It felt strange and good to be back in this place where I had carted home my *Zuckertüte* and played soccer in the fields. The timber-framed buildings and gabled stone houses, the *Marketplatz* and the old Gothic clock tower of the *Stadkirche*: nothing had changed . . . at least outwardly.

I met up with my boyhood friends, Rainer and Frieder. We went out to the fields and found a tree we had climbed as children. We rambled up the slopes of the nearby hills, where I snapped a picture of the town below nestled in its green valley under a sky weighed down with low clouds. We wandered down into the old neighborhood and passed by Rainer's house. There was the little gate and pathway that led to the woodshed where we had pondered the mysteries of the female sex.

I was a German again, laughing with German friends in a German town that had stood here for a thousand years.

It was July, the time of the summer festival called *Festspiel* held in the ruins of a medieval monastery in the old center of town. After attending a performance there among the ancient stones and arches, I met up with Rainer at a bar thinking I would take in some German nightlife, maybe hear what was happening on the German or European music scene. I was disappointed. In 1965, the music was all American or English, no matter where you went.

That was the funny thing about it. I thought that I was going to reconnect with my German roots. But there was nothing to connect to. The real turning point came when I hitchhiked north from Bad Hersfeld to visit my Uncle Rudy and my cousins Heinz Peter and Michel. They lived in Lübeck, not far from Hamburg and just northeast of Bremen, where my family had embarked on our voyage. I showed up on their doorstep one rainy morning, suitcase in hand and soaked to the bone.

Uncle Rudy had gone bald on top of his head, and the way his eyes had sunk into pools of dark shadow above a sharp nose and thin down-turned lips reminded me of my father. He looked as conservative and stolid as he always had, and armored in his suit and tie, as formal. It was clear that he had continued to prosper.

Tante Tütti had grown no less proper. Her coiffed hair was pulled back with a headband, and her face had grown creased and puffy with age, but she looked as elegant as always in her knit skirt and matching jacket. My cousins Heinz Peter and Michel had obviously done well for themselves too. Both had followed their father into medicine.

We spent a day or two in Lübeck before going to a small chalet my uncle owned in Travemünde, a resort town near the Baltic Sea. We sat in the garden, drinking tea out of china cups. It was all very pleasant, very European. I felt my German identity coming back to me.

Then we returned to Lübeck. It was there that I discovered that I was not German after all, but American.

It didn't happen right away. We reminisced some more about family. We took in a tour of the harbor at Hamburg. We went to a museum where I bought a postcard of Dürer's famous brush drawing of an apostle's hands in prayer. It was the night my cousins took me to a restaurant, though, that later seemed to be the whole reason I had come back to Germany.

My epiphany took place at a seafood place near the water in Hamburg. We ordered beers. We took in the waterfront bustle. The waiter came round to take our food order.

"What will you have?" Heinz Peter asked me.

"I don't know," I said, looking over the menu.

"Give him the Hamburger soup," Heinz Peter said then, before I could speak. "We have an American among us, and he'd like something typical."

His words did not register immediately. Only after I had returned home did they echo in my head. When they did, the question of my identity sat before me like a wall in front of a blind man. I had always thought of myself as German, but now, the more I thought about it, it became clear to me that I wasn't German, not really.

More than that, it became clear to me that I had to be something. I sure as hell wasn't going to be a man without a country.

That day I became an American. I would still have to go through the process, but the matter had been decided. After seeing my cousins, it was a matter of getting back to London, catching my flight to New York, and hooking up with two guys I had met in London who promised me a hitch back to Arizona.

The return trip was uneventful. Crammed in the back seat of a Corvaire with the suitcases, I couldn't see much. Only Washington, DC made an impression on me. It was my first trip to the capital. As I stood before the great marble figure of Lincoln, a deep reverence rose up in me.

I became a naturalized citizen of the United States of America the following year, the first one in the family to do so. But Germany always called. The next time I returned was for my father's funeral in October of 1974, and six months later, Lynne came with me on a trip back to celebrate our first anniversary.

If my visit in 1965 had revealed something profound to me about my identity, this latter trip opened a long-closed vault out of which emerged two specters from Germany's infamous past.

The first one materialized out of a beautiful early May afternoon before we even set foot in Germany. Lynne and I had flown into Amsterdam, where we rented one of those funky little European cars and set off through the winding cobblestoned streets to find our hotel. If you know Amsterdam, that means you have to practically be right on top of the building because they are all row houses, each one glued to its neighbors on either side like books crammed on a shelf. We found the place, parked the car and half dragged-half carried our big American suitcases up the steep flight of front steps. Of course, the quaint, little hotel we had chosen had no elevator. So, once we'd registered, we lugged those bags up God knows how many flights of stairs, shoved them in between the walls and the bed, and then collapsed and took a nap.

When we woke up, we came out into the street, expecting the usual carnival atmosphere of Amsterdam. Only then did the silence strike us. No church bells. No bicycle bells. No traffic sounds or snippets of conversation from passersby. In fact, there were no passersby. The city was a ghost town. We wandered around until we finally found someone who spoke enough English or German to tell us what was going on.

Then, it made sense. We had arrived on the Dutch Memorial Day, the anniversary of the day the Nazi occupation ended. A hell of a time to be a German in Holland. Fresh off the plane with my new wife, in Europe together for the first time, and all around us were reminders of how utterly nasty the Germans had been.

After a few days crisscrossing the canals and gazing reverently at the works of the Dutch Masters, we went on to Germany, heading east to Lübeck to visit Uncle Rudy and Tante Tütti. They were the first relatives Lynne would be subjected to, before being presented to Tante Gu in Fulda and Tante Rulli and Uncle Raimund in Winnenden. We were sitting with Uncle Rudy and Tante Tütti in their fashionable flat finishing a pleasant lunch. Lynne was still trying to reconcile what she knew of my background with my uncle and aunt's obvious affluence. Everything in their tasteful home, from the beautifully upholstered furniture to the Bavarian china to the tailored clothing they wore, was testament to a life long accustomed to the finer things.

Lynne had been following Tante Tütti's tortured English fairly well, but at one point she wrinkled up her nose and frowned, tilting her head as if that would shake some meaning into the words. She looked at me saucer-eyed as if to say: *I didn't hear that right, did I?*

The conversation had turned to the economic downturn in Germany. By 1975, the postwar economic miracle of the *Wirtschafts Wunder* had fizzled, and my aunt and uncle were not happy about it. My aunt's German sounded something like this:

"Zee 'konomie ist bat now, nat like beefore," she said, shaking her head. "Vat Gairmany needz now *ist ein kleine kleine* Hitler," she said. "To make zee 'konomie good agen," she added quickly.

I think I nodded or somehow indicated to Lynne that, yes, she had heard Tante Tütti right. It was the first time any of my relatives had referred to that dark period of German history, let alone mention the name that had become poison in the mouths of most Germans.

When it came to the topic of Nazi Germany, I had learned through my early family life that any questions I had about that period would be met with silence. Events that had taken place during those years, even personal ones, were never spoken of. In the photo albums that had been shipped over with our belongings from Germany in 1957 albums my parents had never shared with their sons and which I would find too painful to look at until well after my mother died—there are few graphic references to that time. There is the faint imprint of the Nazi insignia stamped at the bottom of my father's diploma from dental school in 1939; a photo of my mother's brother, Karl, in his *Luftwaffe* uniform; that photo of Graudenz's main square showing long banners emblazoned with the swastika prominently hanging from balconies, while in the foreground a military band plays.

It was a silence my brothers and I had internalized since before we could speak.

So, to hear Tante Tütti speak that venomous name was a real jaw dropper. It left me speechless, but when I got out to the car with Lynne, I let out what I had not been able to say to my aunt and uncle.

"Boy, that was . . . something," I said, shaking my head in disbelief. "*Ein kleine Hitler?* There's no such thing as a *kleine* Hitler. A Hitler doesn't come small."

I wanted to believe that what Tante Tütti had meant by that comment was simply that Germany needed some of the order that Hitler had brought in the early days, when he had built the Autobahn, when he had rallied the nation after its shameful defeat in World War I and had spoken to the German sense of exceptionalism. I could understand if my aunt and uncle had been swept up in the euphoria of national pride when they were young, thinking, *yes, we were defeated in the First World War; we are a proud nation; we are going to get back at the rest of the world.* But who, in hindsight, knowing of the millions of deaths Hitler had caused, could ever voice any kind of comment like the one my aunt had just made to Lynne and me?

The worst suffering had been inflicted on the Jews and Russians and Eastern Europeans. The Holocaust defies comparisons. I knew from my parents' experience, though, that every family in Germany had been affected, had suffered deaths and destruction and displacement on a scale unimaginable to most Americans. I had seen how my own family had been affected, materially and psychologically, how the impossibility of speaking about that period had grown into a wall of silence that separated my parents from their sons and cut off a whole piece of our common history.

How to reconcile such a national history with your personal experience? When I eventually discovered what the Nazis had done to the Jews, I was sick and dumbstricken that such a thing could have been conceived and carried out, much less by my parents' generation. Their complicity—or complacency—left a deep and lasting legacy of shame for their children. The Nazis and everything they stood for were so reviled in postwar Germany that Germans who came of age afterwards have been extra sensitive to even the faintest suggestion that they found themselves superior. It took half a century for Germans of my generation to be able to even wave a German flag at a sporting event, to feel free to be proud of their identity, to be patriotic without being afraid of being taken for a white supremacist or neo-Nazi. When you've grown up with a heritage like that—when you feel personally connected to the chaos that was let loose in Germany when racism turned into murderous policy—you cringe when you see persecution of other groups.

Certainly that background produced a vigilance in me against the currents of racism and bigotry in my adopted country, one fortified during my coming of age during the political and social turmoil of the 1960s and the Civil Rights Movement. When I drove through the South on my way to my externship in Washington DC in 1968, and saw two drinking fountains against a wall, one with a sign reading WHITES ONLY and the other COLORED, a bitter disgust rose up in my throat. And it rises up today in the current climate of racial tension in the United States. I wonder that so many people can ignore the lesson of Germany's infamy. I hope that most will take the philosopher George Santayana's famous caveat to heart: "Those who cannot remember the past are doomed to repeat it."

After those two throwbacks to the dark side of the German soul, the rest of the trip passed with no further reminders of that period. Lynne and I headed south to my hometown of Bad Hersfeld, arriving between the wheat and barley seasons, when the outlying fields were a sea of waving yellow rape. The street of gabled and timbered row houses where my father's dental practice had been looked the same; the grassy common behind it had not yet been turned into a car park. The half-timbered Fachhaus that housed the tavern at the end of the street, outside of which the maimed World War II veterans had sat in my youth, begging for coins, was still there, too. We stopped in for a bratwurst and a pint of beer. "Germany is great," Lynne quipped, "all rape and *Fachhauses*."

I laughed, glad that my rather proper young wife was not above making a racy little joke. Life is serious enough as it is.

My mother died in 1986, never knowing that in a few short years the impossible would come to pass, that Germany would be reunited. I myself could hardly believe it when the Wall came down. I had not thought it would ever fall, at least not in my lifetime. Kennedy's words from his *"Ich bin ein Berliner"* speech resounded in my memory, as did Ronald Reagan's appeal to Gorbachev to "tear down this wall."

I went to Berlin with Gunther in late1989, when most of the wall was still standing. The scene was like a carnival. Enterprising vendors were doing a brisk business selling East German and Russian soldiers' hats, as well as medals and other curios from the former East German State. We bought a few of these souvenirs and then passed through the break in the wall at the Brandenburg Gate. Walking into the Eastern Zone under the grim guard towers and barbed wire, I felt the reach and weight of history. Seeing for the first time how the East Germans lived, how the water and sewer pipes lay exposed on the streets, and how grey and dismal the cityscape was, I reflected on the serendipitous nature of existence. Had one or two details of my early life been shifted, I too might have been celebrating my new freedoms instead of observing others doing so.

A few years later, Lynne, my brother Rüdiger, and I returned. With the Wall down, we were eager to make our way to the East. First we met up with our cousin Heinz Peter (who had become something of a family historian) and set out to find Lockstedter Lager. The refugee camp had been torn down long before, but we remembered the rooster that had assaulted my brother. Rüdiger marked the occasion by having his picture taken while sitting on the signpost for his birthplace.

After that, we made our way to Güstrow, my mother's hometown and the place where my parents had lived as newlyweds. Traveling over the border that separated the two Germanys was like crossing from a bright summer's day into a dreary twilight. The smell of coal clung to the grey buildings, as it did to the rattletrap cars and the downcast pedestrians in their drab clothing and worn, cheap shoes. The past clung to everything as well. Heinz Peter pointed out the pockmarks from the shelling in World War II, still visible on many of the buildings. We wandered around and eventually found the church where my parents had been married in 1937; Lynne and I stood on the steps mimicking the photo that had been taken of the newlyweds coming out the doors in 1937, my father a dashing figure in waistcoat, tails and top hat; my mother slim and glamorous in her bridal gown.

Just across the street stood my mother's childhood home. I knew that it had been converted into apartments, and I didn't expect it to look as it did in the photos taken before the war, but it was still touching to see it and to remember how lovingly my mother had spoken of it that evening before she died. While we surveyed it from a distance, an old woman emerged, prompting me to venture over. I told her why I was there, and asked if I could walk around the grounds. A sense of loss coupled with the bittersweet continuity of things struck me when I saw that the walnut tree still stood, the one my grandparents had harvested to send us packages of nuts in the 1950s and later when they were no longer allowed to leave the East. The woman then invited me in, guiding me through the halls and explaining the original layout. I half expected to see my mother, as she was in those old photos, skirting a corner ahead of us or darting down a stairway.

Our next stop was the town of Rostock, twenty-five miles north of Güstrow on the Baltic Sea, where my father and his brother—Uncle Rudy—had grown up. There, walking through the narrow, winding streets with glimpses of the blue Baltic beyond the red, pitched roofs, Heinz Peter kicked into gear. Elegant, neatly dressed (I don't think I ever saw him without a tie), knowledgeable—very European—he could have been a professional guide. He led us down a street off the main square, stopping before a solid, nineteenth-century apartment house.

My cousin checked the house number and jabbed at some notes he held in his hand. "This is it," he said, "the house they grew up in. They had the second floor."

The building looked freshly whitewashed, but the front facade was scarred, as all the old walls were, with the marks inflicted by the two wars. Heinz Peter probed and scratched at the decades-old bullet holes. "See here. Dad told me he and your father had to duck below the windows when the fighting came close. That was the first War of course. They were just kids."

"I never heard that story," I said. "My dad never talked much about the time before the war."

"No?" he said. "So I guess you never heard about the suicide pact then either."

"Suicide pact? What are you talking about?"

"Our grandmother. The way I heard it was that they all made a pact with one another—Opa, Oma, my dad, and your father—that each would commit suicide if Germany lost the war. Our fathers were teenagers when the pact was made. Dad said that our grandmother somehow acquired enough poison for all of them. She was the only one who went through with it, though."

I shook my head. "But I saw her headstone in the cemetery at Timmendorfer Strand," I said. "When I went back for my dad's funeral. It had 1945 listed as the year of her death."

Heinz Peter nodded. "She waited twenty years to carry through on it, only bringing herself to do it after Germany's defeat in World War II."

I knew almost nothing about this grandmother, the wife of the proud Prussian army officer turned respectable doctor. Questions swirled in the silent street. What had she felt about it all, I wondered. What had she witnessed?

My cousin continued: "Her ashes aren't there, though, in Timmendorfer Strand. According to her wishes, they were scattered over the North Sea."

"That's another thing I never knew," I said. "And where is our grandfather anyway? There's no stone for him there."

"You know," he said, screwing up his face. "That's a question I haven't got an answer for."

The personal is political, the feminists said back in the 1970s. The reverse is also true. The fall of the Berlin Wall was a deeply personal event for me. But once I had seen the Wall, I still felt no resolution about it. It was as if I had imagined its dismantling. I couldn't reconcile the reality that it was gone with the old reality of its existence over fifty years of my life. I think I needed some way to express the joy and gratitude I felt over the reunification of Germany before I could absorb it all.

I returned to Berlin in 1990 and chiseled off a small chunk of the Wall, one of a steady stream of visitors intent on lugging away a piece of history. In 1992, Lynne saw it when we went over for the tour of my parents' hometowns. A decade later, in 2003, we brought Alissa and her future husband Tyler, Christopher, and Stephen over with us when we traveled to Germany on a search for venues in Berlin and Dresden for the combined convention of German and American neurosurgeons the following year.

Finally, in 2004, while we were in Berlin for the first part of that convention, Lynne and I were invited to one of my German colleague's homes for a private piano recital. Afterwards, along with our hosts and the twenty or so other guests, we walked to a nearby restaurant for a late supper. It was a cozy place, and our party took up every table but one that had been reserved.

About an hour later, the mystery party arrived. It was the former chancellor, Helmut Kohl, now a white-haired man in his mid-seventies but still ruddy complexioned and hale. His dark eyes looked all the sharper for his remarkable eyebrows, twin smudges that, unlike his hair, had refused to pale. He walked in with a few companions and sat at the table. Eager to take advantage of such an opportunity, our German host asked the Chancellor if he would meet with the group of neurosurgeons. No, Kohl said, he couldn't do that, but he would be happy to meet with the president of the organization.

That was me.

A photograph of that meeting hangs in my upstairs office at home. There I am with the man responsible for the reunification of Germany. I talked to him a bit that night, and thanked him for what he had done.

I presented that talk in Dresden in a tasteful but nondescript convention hall designed for such gatherings. It could have been New York or Tokyo or Hong Kong. That evening, though, the members of the German and American associations met for a formal dinner in the banquet room of one those domed and gabled and turreted castles that had made Dresden a pearl of European architecture. As with many of the Baroque and classical buildings that were pulverized in the epically destructive firebombing of the city at the end of the War, only the facades of the structure were original, but unlike those buildings whose interiors had been renovated in a sleek, clean style, great care had been taken to recreate the opulent interior of this particular castle. The high, arched ceilings receded into shadows. A thousand candles flickered in chandeliers and wall niches.

Amidst all that beauty and elegance, all that evidence of the genius and artistry of the culture into which I had been born, what left an abiding effect on me was, again, the feeling of camaraderie—even solidarity—that my talk had evoked and that still lingered in the air. It was as if a communal voice were whispering: *We're all right as Germans…We are not such awful people… We welcome you Americans here… We are all okay now.*

A few days later, I boarded the plane for home. I felt that I had come full circle at last, and had finally laid the past to rest.

CHAPTER 24

THE GURU OF SPINE

Try not to become a man of success, but rather a man of value.
ALBERT EINSTEIN

FIVE YEARS BEFORE our stay there, on November 26, 2008, smoke had billowed from the pointed arched windows of the ground floor, obscuring the Moorish-Byzantine facade and mushrooming up to engulf the red onion dome and turrets of the century-old hotel. Friends had immediately reminded us of this terrorist attack when I announced my plans to accept an offer to deliver an important address in India, but once there, the incident had slipped from our minds. Lynne and I passed under the gracefully curved and pointed arch and up the marble steps, nodded to the doorman—a turbaned figure in white leggings and knee-length tunic from the days of the British Raj—and moved through the rosewood doors of the Taj Mahal Palace Hotel in Mumbai. The vaulted alabaster ceilings, the onyx columns; the grand cantilever stairway with its plush carpeting in deep carnelian red, and the view of the galleried corridors above: it knocked my socks off.

The hotel was a stark contrast from the drive in from the airport. Even though I'd had three days of cultural transition in New Delhi, I had to shake my head clear of the images embedded in my brain: the motorway like a massive exhaust-clogged construction site; bedraggled palm trees choking for air amid bus stops and store-fronts; our taxi darting around tuck-tucking, motorized rickshaws, bicycle rickshaws, and vendors pushing carts, while occasional women in saris or long tunics and headscarves flitted across the street like colorful butterflies. Then the city center, with its towering apartment blocks and congestion; its makeshift dental and barbershops servicing clients on the sidewalks; and maimed beggars—many children—beseeching passersby with dusty hands (or stumps). All the while horns shrieked and beep-beeped; engines rumbled; the oppressive heat rose up and bore down; and the smell of petrol fumes, dust, and organic matter assaulted our American noses.

Stepping into the cool, stately Taj Mahal Palace, I felt the refuge of it—and my privilege—acutely.

I had received the invitation to deliver the prestigious 2013 Ginde Oration at Bombay Hospital in July of 2012. I would be following in the footsteps of the "illustrious" neurosurgeons who had attended every year since 1992 to speak and perform surgery in different clinical conditions. My predecessors included M. Gazo Yasargil, the inventor of the Yasargil aneurysm clip; Don Long, who had been a finalist for the queen's surgery; and Robert Spetzler, among others. Given my friendly rivalry with Robert, I couldn't help but crow to myself when I nonchalantly told him that I'd been invited. There was one problem, however, of which I informed my correspondent—a senior, highly respected neurosurgeon named Dr. Sanat Bhagwati—before officially accepting: I had stopped operating in 2010. Not a problem, he replied. Might I consider doing a cadaveric workshop apart from my lecture and symposium? I agreed and on Sunday, April 20, 2013, I entered the large, white, mid-twentieth-century block of Bombay Hospital with my guide and colleague, Dr. Chandrashekhar Deopujari.

I have given 162 talks in 28 countries over the last five years. In 2014 alone, I lectured in Holland, New Zealand, Jamaica, Austria and the Czech Republic, as well as in multiple cities in the United States. I have to say, India was a uniquely gratifying—even head-swelling—experience. Dr. Deopujari was nothing short of grandiloquent in his praise. "Come, come, Dr. Sonntag," he said, ushering me around with great enthusiasm during tea and registration. "My colleagues are eager to meet you." Then, stopping before small clusters of guests, he sang the same glowing praise: "I present to you our Ginde orator, the estimable Dr. Sonntag. He is the "guru of spine."' By the end of the trip, I think I heard that phrase at least a dozen times. The hyperbole was rather embarrassing, but appreciated all the same.

After a short, moderated "Meet the Professor" session, I gave a couple of short talks and responded to other doctors' brief presentations on topics such as "A rare case of Hemangioendothelioma of Dorsolumbar Spine" and "Disc space infection following Endoscopic Discectomy." For understandable reasons, Lynne did not attend these presentations with me. But she did appear for my 10:30 "oration" on "Cervical Spine

Instrumentation: Past, Present, and Future," after which we toured the different departments of the hospital.

The tour was another stark contrast, both to the resources I enjoyed at Barrow and the lavish service Lynne and I received at the hotel, where a personal houseboy anticipated our slightest whims, placing bookmarks in our books and leaving us lovely pens. Here the only wheelchairs we saw were two decrepit models slumped in a corner, like props in a Hitchcock film. Aides folded laundry on the worn linoleum floor. In the pediatric ward, two nun-like nurses in white ministered to two or three dozen children of all ages confined to institutional iron cribs, while family members gathered round—the lucky ones sitting on wobbly plastic chairs and the others squatting on the floor. The ICU was particularly alarming; I brusquely told staff to elevate a patient's head before his prone position killed him or otherwise harmed him. This is not to say that the problems I perceived were due to laxness on the part of the proud, opinionated and dedicated doctors I met but to what plagued the country in general: lack of resources.

The morning lecture and talks took place in the S.P. Jain Auditorium at Bombay Hospital, a conventionally comfortable meeting room in unassuming, tawny tones. For the cadaveric workshop, however, we moved to the venerable Grant Medical College established under the British in 1845. Though the institution had expanded since its foundation, the campus now being a mix of newer and older buildings, the Colonial architecture of the original building evoked the century of its inception.

The lab itself was newer, dating from the 1950s or 60s. A very basic space—whitewashed walls, large windows, and ten simple cadaver stations—it was the kind of room you've seen in photographs of mid-twentieth-century American hospitals. Still, like most of the facility, it had no movement of air. The pickle-like odor of formaldehyde permeated the room, taking on an unusual potency in the moist Indian heat.

I took my place in the front of the lab. Before me lay my cadaver, face up, its cervical spine already dissected down to the bone. The forty-odd Indian neurosurgeons pressed in to observe, along with a sprinkling of radiologists and other attending doctors. As a sign of respect, my hosts had given me a set of brand new scrubs, over which I wore a heavy lead apron, a thyroid shield, gloves, gown, and my customary full-head surgical bonnet. Crisp as toast when I'd put them on, I could feel the scrubs wilting in the close, hot lab. It was two o'clock in the afternoon. The temperature was in the 90s. The humidity was in the high 80s.

I began the demonstration with instrumentation of the anterior cervical spine. As with all surgery, foundational to this is a precise and detailed grasp of anatomy. I indicated with my knife how the soft tissue had been dissected in order to find the normal plane between the carotid lateral and the esophagus and tracheal medially, driving home this major point. "The placement of the screws is critical," I told the group, "in order to avoid inadvertently damaging the spinal cord, the nerve roots or the vertebral artery. Consequently, it is very important to use anatomical landmarks to determine the midline of the cervical spine. These landmarks are the sternum and the nose, and in line between the sternum and the nose, which is the midline."

The doctors pressed forward as I identified further landmarks: a spot between the two longus coli muscles—the muscles along the cervical spine that variously bend the neck forward and flex the head—and the uncus joints, the dish-shaped upper and lower bone lips of C3, C4, C5, C6 and C7 that keep the discs separate from one another. At some point I also raised the fluoroscopy screen, which sits ready at the side of the cadaver table, allowing the doctors to view the procedure in better detail.

After ten minutes or so on the front, we turned the cadaver over. Again I started with a plate, this time the mass lateral plate, which is only used posteriorly. I didn't think of it then, but now I can note with both irony and satisfaction that this is the device that propelled me into controversy back in 1988, when my orthopedic colleague walked out on my first spine surgery using the then cutting-edge technique. Thirty years later, it had become standard. And along the way, I had moved from a pariah among orthopedic surgeons to Dr. Deopujari's "guru of spine."

The plates were just the beginning, however. The main thing I was in Mumbai to demonstrate was screw fixation, the first technique being lateral mass screw fixation, which, since the 1980s, had developed apace with the plate as a method in stabilizing the cervical spine. In this technique fixation is achieved with two screws 16-18 mm long placed posteriorly at each level of the spine that needs stabilization.

Like so many spinal techniques, screw fixation is fantastically difficult. The starting point of each hole you drill has a recommended angulation that varies depending on the particular vertebra you are working on. You're placing screws medial to a certain point in the lateral mass, then aiming the screw 30 degrees rostrally and approximately 15–30 degrees laterally. And that's just if you are using lateral mass screws. An alternative way to fixate C3 to C7 is to place pedicle screws into each vertebra, but those entry points vary as well. It's not as if you have a diagram to follow either, of course. You have to internalize this stuff. Then it's just you, your patient, and their bloody, unique, frequently surprising spine.

After the lateral mass plate and screw placement, we moved on to my finale, C1-C2 screw fixation. This was the procedure I had made my calling. The primary stabilizing element of this relatively new and surgically challenging technique—known as the Magerl technique, after the surgeon who devised it—are two long trans-articular screws bored posteriorly at an angle through C2 and up into C1. While, once the screws are passed, immediate fixation is attained, as with the other techniques, the screws run very close to a vertebral artery, so great care must be taken with the positioning.

One more time I demonstrated the sequence of steps: the dissection of the lateral mass of C2; aiming the pilot hole to the tubercle of C1 through the pars of C2, and entering the lateral mass of C1; and so forth. Then, the many variables a surgeon encounters: perhaps the pars is too small to be able to place a 3–4-mm screw; maybe the lamina arch at C1 is too small; you might need to depress the C2 nerve root to place the screws in the lateral mass of C1, etc. All this I had struggled with myself. All this I had mastered. It is for this reason of course, that well seasoned surgeons are most in demand to teach such material—indeed outside of the United States we are treated like rock stars.

And so it went. I followed up with some other procedures: more do's and don'ts of putting anterior plates on, a fluoroscopically controlled reduction maneuver and rod fixation, and so on. Then it was three or four hours rotating among the cadaver stations. Sweat trickling down my face, I guided the surgeons in their turn, answered their questions, and repeated the techniques when necessary. By then, hours into the workshop, the odor of forty sweating humans and ten cadavers, mixed with the cat-urine smell of the formaldehyde, was overpowering—barring that teamsters job in a Phoenix warehouse the summer before starting med school, it may have been the most stifling environment I've endured. I think my shoes had puddles inside by the time I was done. When I finally went to change, I looked around in vain for a towel, and ended up air-drying my body as best I could in the close, humid quarters before taking a taxi back to the hotel in my street clothes.

The workshop that day was typical of the talks that it is now my privilege to give all over the world on the development of spinal neurosurgery, charting and teaching the techniques that have paralleled my career. When Steve Papadopoulos had said I was a spear already in flight in 1988, I am not sure we could have envisioned the explosive growth of the spinal implant industry. From early wiring and bone fusions, to the crude rod I had used on TJ in 1989 (a procedure duplicated on Christopher Reeve in 1996 following his horse-riding accident), to plates and screws, rod systems, pedicle and transarticular screws, and interbody cages, the climate has been one of continual experimentation and technological development.

At the same time, diagnostic imaging of the spine has taken huge strides. Static X-rays had been around since the turn of the last century, followed by an early form of fluoroscopy—the direct live imaging of the patient's anatomy. The next major leap was the advent of computed tomography (CT) in the mid 1970s, a technique that brilliantly defines the bony skeleton and spinal canal and has greatly improved with time. After that came the incredible boost of magnetic resonance imaging (MRI) in 1977. The introduction of intraoperative imaging using MRI and CT technology followed.

One example of this is stereotaxy—or image guided surgery. With this development CT and MRI can be used in real time to define anatomic locations with pinpoint accuracy. However, while stereotaxy became common in cranial surgery by the 2000s, problems with the intraoperative registration of the bony spine have still made it unavailable in spinal surgery. Fortunately the much older technology of continuous fluoroscopy—first developed in 1895 and still essential for me in India in 2013—continues to provide direct live imaging of the patient's spinal anatomy.

What has all this meant for patients? According to federal figures, the number of spinal fusions rose from 56,000 in the United States in 1994 to 465,000 in 2011 (a six-fold rise). Moreover, whether undergoing spinal fusion or not, according to the National Spinal Cord Injury Association, as many as 450,000 people in the U.S. are living with a spinal cord injury. Though of late there has been a necessary scrutiny of the readiness of some spine surgeons to recommend surgery over other treatments, for half a million people—in the United States alone—advances in spinal surgery hold the promise of restored function and a return to normal activity.

Such promise is perhaps what propels me to continue traveling around the world to teach six years after my retirement. There is also the sense of cementing and furthering my main contribution to neurosurgery. Steve Papadopoulos put it this way when I recently ran a question by him.

I had popped into Steve's office for his take on the topic of another presentation I was scheduled to give in the coming spring for The Joint Section on Spine. "I'll be speaking as honorary mentor or teacher," I told him. "You know, the 'Old Fart' presentation. They want me to talk about what my contributions to spine surgery have been. What do you think?"

Steve nodded and after a moment gave me his answer. "You know, Volker," he said, "the plate design and fusion techniques you were involved in engineering; the wiring and the patents . . . all that was fine and good. But the real contribution you made was the fight to have neurosurgeons do the spine. That was the profound change you helped bring about."

I think Steve was right. My biggest achievement wasn't all those individual advances but the very establishment of spinal neurosurgery as a subspecialty in its own right. Now, having helped to birth that baby, I plan to nurture it as long as I am able.

Of course there is also the pure joy of teaching. I still marvel at my good fortune in finding this out about myself in 1978 when I started the board review rounds. My experience over nearly four decades has driven home to me what a personal travesty it would have been had I not followed the academic side of my profession. As early as 1979 the residents voted me Teacher of the Year at Barrow, an honor I was proud to earn several times during my career there. Then in 2000 the residents voted both Robert and me "Mentors of the Millennium." One of my proudest teaching moments came in 2006, when I was named one of ten annual recipients (and the first neurosurgeon) of the Parker J. Palmer Courage to Teach award. Established by the Accreditation Council on Graduate Education (ACGME) in recognition of the crucial role of teaching in the formation of medical professionals, the award is based on an influential book called *The Courage to Teach* by Parker J. Palmer, a senior leader in the field of education. It was a great honor to receive that award, though teaching has been a reward in itself. As the great seventeenth-century philosopher Baruch Spinoza said, "Happiness is not the reward of virtue, but virtue itself."

Then there is, finally, the matter of paying back some small portion of what I have received. Teaching is part of the bargain I struck forty years ago, when I was the recipient of knowledge accumulated and passed on over centuries—and not just in med school and my residency. I remember the long days of tutelage under Sandy Larson in 1988, his tremendous patience in teaching me and the other doctors the intricacies of rods and hooks. Like Dr. Larson, I have been a link in a long and noble chain that stretched from the ancient Greek physician Hippocrates—who said, in his book, *On Joints*, that physicians "should first get a knowledge of the structure of the spine, for this is also

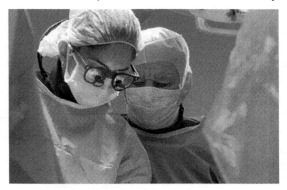

Training new generations of neurosurgeons has been a supremely rewarding aspect of my career. Dr. Rasha Germain, shown here, was a Barrow resident at the time.
Photograph by Gary Armstrong, used with permission from Barrow Neurological Institute.

requisite for many diseases"—to the father of modern neurosurgery, Harvey Cushing and his pioneering British contemporary, Victor Horsley, who performed the first laminectomy for spinal neoplasm. In my own career I have seen that chain grow into a global network of shared knowledge. Nothing makes me prouder than being a link in that history.

I came out of the cadaver lab that evening a steamed dumpling and returned to the Taj Mahal Hotel. Fortunately it was only a five-minute taxi ride; I think it took me longer to navigate the grand lobby, halls, and elevators to my room than it did to get back. Lynne and I had agreed to meet at the pool. I changed out of my street clothes into my swimming trunks, threw on a snowy robe, and made my way down to the ground floor. I wanted to get in a few laps before Lynne and I were expected at eight for dinner on the lawn of the Willingdon Club—(a former English gentleman's club that, according to our host, had until recently kept two historical signs in place at different entrances: one for "Whites Only" and the other for "Indians and Dogs.")

Now that the main part of my work was done, I could relax. I crossed through the lobby, zigzagged around the low tables in an elegant lounge where stylish, international guests sipped cocktails, and came out on the other side to an open-air, arched colonnade lined with potted palms— very "Jewel of India." Beyond was the pool. I sauntered into the expansive courtyard between the two wings of the hotel and saw Lynne waving at me from a poolside chaise longue. The sun had dipped behind the massive central building, giving the air a golden glow that belied the street noise and exhaust behind a nearby wall. The long oval pool was a crystal aquamarine jewel.

"Glad you made it," Lynne said when I reached her. "We'll have to dress quickly for dinner, but it would be a shame to miss the pool. It's wonderful."

"Me too," I said, untying my robe and thinking how beautiful Lynne looked with that big smile on her face.

No sooner had I let the robe drop, however, than Lynne broke into peals of laughter.

"What? What?" I said.

"Volker, look at yourself. You're all blue!"

I looked down at my torso and legs. Sure enough, I looked like I'd been dipped in ink. I had been dyed a deep indigo blue from those brand new scrubs.

THE MENSCH

It is not the honor you take with you, but the heritage you leave behind.
BRANCH RICKEY

FROM THE CULTURE into which I was born, I learned the ideal of the *Mensch. Mensch* is a German and Yiddish word derived from Middle High German meaning a person of integrity and honor; an upstanding, admirable person; a person possessing such characteristics as fortitude and firmness of purpose. If you are a *Mensch*, you are a complete person. If you are a Mensch, you are a human being.

I believe that the road to happiness in life is the road of the *Mensch*. I believe it is the ideal of the *Mensch* that led to the development of my philosophy of the Four Hs (hope, hard work, honesty, and humility). I believe that all that I now cherish—my beautiful family; clarity of mind and a strong body; and the extraordinary privilege of being a physician, of caring for people in their time of need—have been a result of trying to follow the road of the *Mensch*.

I think now of my beginnings. I picture my mother with me in her arms and Gunther in hand fleeing through the desolate ruins of Graudenz that cold January many decades ago. Who could have told her that the infant she cradled would grow up to follow the path into medicine that his father and grandfather had paved; indeed, that he would grow up to become a neurosurgeon? Who could have assured her that her child would even survive?

There were struggles along the way: the refugee camp; immigration; setbacks and accidents and exhaustion and failure. There was acknowledging and somehow redeeming the sacrifices that my parents had made. When they rose before me, some of those obstacles seemed insurmountable. Now I see that each obstacle was both a hurdle and a gift. Each obstacle moved me closer to the inestimable joy that lay not in the material rewards of my profession but in a life spent caring for people.

Ralph Waldo Emerson said: "Life is not a destination but a journey." My journey has been the mysterious, exciting, enlightening, and, yes, difficult one of the physician. It has been a journey during which many teachers have emerged, among them the human body itself, the body that even in death—as a cadaver—was gracious enough to teach me about life, about the details of human anatomy and physiology.

It has been a journey of profound emotions, of the sadness and disappointment that accompany pain and suffering and loss of life, but also of the joy and amazement that come from ministering to a fellow human being and aiding in that individual's recovery.

It has been a journey of the greatest intimacy with others, a journey that gave me privy not only to patients' secrets and stories and private lives but even to their bodies and being, in exchange for the hope that I would be able to alleviate their pain and restore them to a healthy life.

On this journey, I assumed the doctor's role of shepherd and guide, helping patients navigate unfamiliar terrain and make informed, critical decisions, even when that decision was the heartbreaking one of not subjecting a loved one to further intensive medical or surgical care, of allowing normal passage from life to occur.

I feel enormously blessed to have been a medical doctor, for only an MD works on the brink of science and humanity; only an MD integrates art, science, and research to touch human beings in such a profound way. Only medical doctors are given the responsibility—and the trust— to apply their knowledge and discoveries from research so directly and intimately to a fellow human being. Donning the white coat of my profession gave me the satisfaction of helping people in a way that no other profession could have offered me.

There have been headaches to deal with, and those headaches will remain: bureaucracy and health care reform that even now is changing the way healthcare is paid for and how a patient receives that care, and soon may even dictate which patient a doctor can or cannot take care of. But underneath, and in contrast to the turmoil of the political-legal world, the core principle of the Hippocratic oath beats like a heart. The professional and ethical standards it professes allow (indeed require) the physician to transcend politics and the meddling of policy makers. Whatever storm of controversy might be swirling around the medical community, whatever the current terminology, the overarching objectives in the operating room or examining ward or at the patient's bedside remain clear and simple: HEAL THE SICK; EASE THEIR PAIN; RE-SPECT AND COMFORT THE PERSON.

Given the fundamental nature of the task doctors have been charged with since the Hippocratic oath was penned in antiquity, I don't believe the patient-doctor relationship will or should change, despite changes in technology and advances in diagnostics, genetics, research, biomechanics, pharmacology, robotics, patient care, surgical techniques and more. Armed with knowledge and sustained by humility, compassion, and a commitment to treat all with dignity, the physician will continue to fill the role of teacher, counselor, and healer; and science, art, and the magic of medicine will continue to underlie the relationship between the doctor, the patient, and that patient's family.

And that brings me back to my own family.

My daughter Alissa (*"die schöne, elegant* Alissa"* my Tante Rulli always said) recently gave birth to Lynne's and my second grandchild, a sister named Annalise for our grandson Tucker. It seems like yesterday that our daughter was dancing around in a pink ballerina costume at Halloween, or belting out "Great Balls of Fire" at a talent show, or winning the state tennis doubles title in high school. Now, having graduated from college and experienced the professional world, she is a wife and mother, raising a family of her own, creating something new but also carrying on the love and traditions that have been passed on to her as her birthright.

My sons are testing their mettle in the great world too. I look at Christopher, now a successful business owner, and remember a small boy muted by his vivacious sister, a child his kindergarten teacher later suggested we have evaluated by a speech therapist. It turned out he was simply speaking English with my German accent. I still see in him, too, the focused young athlete he was at ten. I was proud of his skill in athletics—little league and soccer, later track and lacrosse—but even prouder of his quiet dignity and humility when he won. It was only during his high school years that Lynne and I discovered what a killer wit and sense of humor he has.

And Stephen, our youngest, now two years out of ASU and into a career selling commercial real estate. So much younger than Alissa and

Chris, he had four parents to teach him, and it served him well. Lynne remembers him telling some story to Robert when he was around four, upon which Robert said, "You have just said more words to me than your brother has said in his whole life." Bright and competent with a good mind for math, he is a real people-person and an excellent athlete in his own right. Most endearingly, he has always been willing to hang out with his father—even during his high school years. As you might imagine, that just tickles the socks off me.

How gratifying it is to watch your children grow up, to see not only the characteristics bequeathed to them by their lineage but also to witness the emergence of unique and amazing people in their own right. Alissa—fair and slender and vivacious, passionate about yoga—is clearly her mother's daughter. She is also the embodiment of a philosophy passed on to me by my own parents: "Stay healthy and strong," my mother had said to me before she died. But Alissa is something entirely new, too, as her brothers are, something Lynne and I could never have envisioned.

There is continuity and there is change.

I consider my greatest accomplishment to be my family. And conversely, my family has been my greatest support. In the office I still keep at BNI, the lyrics to the song "Wind Beneath My Wings" are tacked up over my desk. Alissa sent me those lyrics, along with a touching note about my being the "wind beneath her wings." I hope she realizes that the reverse is true, that she and Chris and Stephen, and especially Lynne, are the wind beneath my wings.

I think back now to a day long ago, when I was working at Mr. Reynolds's chicken farm. I had finished cleaning out the troughs that day. I was tired, dirty, discouraged. I stopped for a moment and took a small oath. If I have children someday, I swore, I am going to make sure they will never have to do this kind of work.

I feel blessed that I was able to keep that oath, blessed that I was able to provide for my children and that my children have used what their mother and I provided to good advantage.

And I remain convinced, more than ever, that if a person is honest and works hard—if a person strives to be a Mensch—things do have a way of working out.

ACKNOWLEDGEMENTS

FIRST AND FOREMOST, I want to express deep gratitude to my loving wife Lynne, who has stood by my side and been the rock on which I have leaned for the last forty-two years. Her support and contributions were invaluable as I worked through organizing my memories.

I want to thank my three children, Alissa, Christopher, and Stephen, who are such a big part of my story and who provided feedback and encouragement every step of the way.

Thanks are due, too, to Robert F. Spetzler, dear friend, colleague, and Chairman of BNI (whose guidance, friendship and leadership I cherish); to Doctors Bill Buchsbaum (who inspired me to go into neurosurgery), Bennet Stein (under whom I did my residency), Steve Papadopoulos, Nick Theodore, Curtis Dickman, and my other colleagues, as well as the fellows and residents at Barrow Neurological Institute with whom it has been my privilege to work; and to then senior editor at Barrow Publications, Shelley Kick, my assistant Debbie Nagelhout, my then-secretary Cheri Ebert, and the creative staff at Barrow Neuroscience Publications for their invaluable contributions and guidance in making this book a reality.

Early readers and advocates for the book deserve recognition: Sr. Sara Marie Belisle, OSF, whose assessment and praise for an early version of the manuscript was a great boost; Tom Lombardo, whose astute comments improved the story; and Kathleen Papajohn, who was instrumental in bringing the first phase of this project to fruition. Thanks, too, to the late Dr. Ed Sylvester, whose book, *The Healing Blade*, served as a useful historical resource, and to Carol Ackerman, Suzanne Baars, Dr. Allan J. Hamilton, President of ASU Michael Crow, and Dr. Sanjay Gupta for reading the manuscript and providing comments.

Three others deserve special recognition and gratitude: my agent Claire Gerus, who identified the essence of the present book in the first version of the manuscript and guided the drafting of the rewrite with patience and insight; Lisa Hagan, for seeing the book into print; and Suzanne Chudnoff, for her invaluable review of the final manuscript.

Finally, I owe thanks to Jeanne Lombardo for her talents, patience, persistence, guidance, and help in writing my story...the story of the American Dream.

Made in the USA
San Bernardino, CA
02 November 2019

59330174R20146